TROPPO
ARCHITECTURE FOR THE TOP END

PESARO PUBLISHING

CONTENTS

TROPPO: ARCHITECTURE FOR THE TOP END

Introduction by Glenn Murcutt	7
Chapter 1 **GOING TROPPO**	11
Chapter 2 **TROPPOVILLE**	21
Chapter 3 **NEW CHALLENGES**	41
Chapter 4 **NGAD ARRI BOLKNAHNAN**	55
Chapter 5 **PRODUCTIVE AFFLICTION**	67
Chapter 6 **TOWARDS A TROPPO ARCHITECTURE**	83
Summation **IMPENDING PERMANENCE**	87

TOM HARRIS DONGA, HUMPTY DOO 1990

INTRODUCTION

Glenn Murcutt

On my first visit to Darwin, in the mid 1980s to advise a Government Department on tropical housing for the new Palmerston township, I happened to be walking one evening through Central Mall. In a shopfront office, clearly one belonging to a firm of architects, I noticed some of my own work - the Mt Irvine and Kempsey houses - pinned on a wall. It was a real surprise. In the office the next day, I received a phone call from an architect called Phil Harris. Somehow he and Adrian Welke had found out I was in town. Phil explained that my work had meant a great deal to them. I made the connection to the night before. They were Troppo Architects. I knew who they were, but I didn't know what they were doing as architects. I soon found out.

After buffalo steaks for dinner, I realised I'd met two very special architects: generous in attitude and spirit. It was instant compatibility. Later I spoke with a local architect who said that the work of these two young turks from Adelaide was frankly rubbish. I disagreed vehemently. Having grown up in New Guinea and knowing the work of my father who understood the importance of lightweight elevated buildings and cross-ventilation, I said to him 'these guys are on the right path'.

I became good friends with Troppo and with their families. On subsequent visits to Darwin, I saw their work. The Green Can was clearly their beginning. In that job lay the founding principles for a future practice of tropical architecture. Troppo also showed me the early 20th century slatted houses, and the 1930s houses of Beni Burnett. The Burnett houses reminded me so much of the government pre-war work in Papua New Guinea, in Bulolo and Lae. The slats gave privacy, ventilation and shade, but from within one could see out. The windows that hinged from the top reduced the level of sunlight on the wall and also reduced glare when looking out. It was Troppo's strong interest in this direct architectural response to climate that I found compelling about their practice. Even their sketches from student days showed the direction in which they would head. It was a path totally different from that followed in the south.

Such was the importance of Troppo that I encouraged them to think seriously about joining the NT Chapter of the Royal Australian Institute of Architects. It was typical of Troppo not to have joined. Their whole philosophy may have seemed renegade, outrageous even to some, but to me, it was Troppo who were focussed on an authentic architecture for the tropics. It was most of the other architects in Darwin who were so unresponsive to their special place. I felt Troppo needed to have greater influence. They did join the RAIA, and won the Tracy Award the very next year, but it was the national award that was a special sort of recognition. It was acclaim in the broadest sense. And this was important to them, to the profession, to Darwin and to the public. To use the RAIA as a platform for support was one way of bringing their work to public attention, one way to influence.

Troppo also introduced me to Kakadu and the beauty of its landscape. I saw their work there and was impressed. When they asked me to collaborate with them on the Bowali Visitors' Centre, I hesitated, because of my own workload. But after much discussion with Adrian and Phil, they convinced me to put in a joint submission. When we were awarded the job, it was an extraordinary experience. Here were two architects completely on side. The whole process seemed to be so fast and yet so intensely productive. We studied rock shelters, rainfall and cyclonic conditions, wind patterns, temperature, and humidity. We also examined the escarpments and the cross-sections of caves, where rock paintings were applied to cave walls, the different levels of prospect and refuge, how water was to be collected, how sites were to be entered from the side, and how to respect the guardian spirit of the place. We soon understood that the project was more about connections with the landscape and path, than a series of spaces. We worked flat out. Then just as we had confirmed our planning strategy for the site, Adrian began to draw up the project while we were still finishing the design!

During documentation, Phil and Adrian sent drawings to me in Sydney by air-bag. I learned a lot in the process. Building practice was a lot rougher than I was used to. My suggestions for increased levels of refinement to the building were not necessary. Troppo were right. The construction costs were prohibitive and my predilection for finesse was just not relevant here. It wasn't the place for it. In this, Phil and Adrian clearly understood their place, their client and their way of building, and the budget. Troppo were incredibly adaptable, not precious, adventurous and committed to the project.

My appreciation for Troppo has continued to grow. They are devoted to Darwin. They arrived there without any conventional architectural baggage. In other words, they came with their eyes open, and with a background to see. They also found what they hoped they would find - the principles of an architecture appropriate for the tropics, and a spirit of place.

CHAPTER ONE
GOING TROPPO

Historically, the phrase 'going troppo' came into prominence during World War II. It meant to be 'heat affected', one went 'off', a little mad in the heat of the tropics. This is the story of two architects who 'went troppo', but in doing so, found another state. For them it meant becoming acclimatised, not going 'off' at all, but understanding the place. 'Going troppo' was revealing - even enlightening _ and it would provide the basis for a new practice of architecture in the Top End of Australia.

Too often, the profession of architecture is overly earnest, rarefied and painfully self-conscious. The story of Troppo Architects is an unusual one in the history of Australian architecture. It has its comic, tongue-in-cheek side. Humour is an intrinsic element of the practice begun by Adrian Welke and Phil Harris, but it also has an intensely serious component. Their work, between 1979 and 1999, has recovered for architecture in the Northern Territory an idea of climate, region, historic context, and formal variety. An idea of professional and aesthetic respect, and above all, an immensely appealing idea of how to live and work in the Top End's demanding seasons of the 'Dry' and the 'Wet'. This is a place that lies midway between the Equator and the Tropic of Capricorn where author Xavier Herbert has said 'nature... looms larger than man'. Troppo have given a lot back to the place - their work is not like that found in any other Australian state. It demonstrates that there are many different and valid architecture cultures across the continent. With their own special degree of self-deprecation and irony, and their willingness to be part of a unique place, Troppo have made the Top End their own.

Early days

At the end of 1977, four architecture students at the University of Adelaide wanted to do something a little different for their pre-final year project. Phil Harris, James Hayter, Justin Hill, and Adrian Welke decided that they would travel around Australia for four months. They had been inspired by an article in *Architecture Australia* (April 1975) where Sydney architecture students Paul Pholeros, Phil Rose, Wal Zagoridis and his wife Irene had described their travels around Australia in their "Australian Communications Capsule", a converted double-decker bus. Instead of choosing a bus, the four gangly Adelaide youths, in shorts and calling themselves ACME ANYWHERE, set off in a Volkswagen Kombi. They followed the coast west from Adelaide through Eucla, Albany, Perth, Geraldton, Port Hedland, Broome, Darwin, Normanton, Cairns, inland to Winton, back to Rockhampton, Brisbane, and from Sydney they made their way inland back to Adelaide. When they returned in early 1978, they wrote up their experiences and observations in a joint research report entitled "Influences in Regional Architecture" (May 1978) and publicised their trip with a three projector slide show with sound track highlighting regionally appropriate architecture from all over Australia.

The Top End

The trip was intended as research for architectural identity through an association with region. Determinants such as climate, availability of materials, as well as social, economic and technological conditions formed part of the investigation. As students, the ACME team had been impressed with lecturer Stefan Pikusa's housing studies, but more particularly their quest for discovering regional identity arose through the sheer lack of documented Australian architectural history. JM Freeland's *Architecture in Australia* (1968) was the only major text available to students, and was woefully inadequate in terms of describing buildings and places other than those located in the capital cities of southeast Australia.

Of all the places the four students visited on their tour, Darwin and the Top End were the most revealing. The Top End is the northern third of Northern Territory, the area of Australia closest to Indonesia, closer to the Equator than Bangkok, and it has a tropical climate. Invariably described as 'hot humid', the region is susceptible to the influence of the north-west monsoons as well as the dry heat of Australia's arid centre. The 'Wet' season runs from November to April and the 'Dry' from May to October, with minimal seasonal variation from a daily maximum of 32-34C. The ACME team were struck by the region's isolation, its poor agricultural land and its lack of rural town centres. A sparse population meant shortages of labour and market sources for any sustained commercial and industrial development. Through their research they found that the development influences on the Northern Territory had invariably been from somewhere else. From the first isolated settlement at Melville Island (1824), to the "South Australian pastoralists and explorers, Queensland cattlemen, Asian 'coolies', multi-national mining companies and Canberra public servants", the Top End had been the subject of a myriad of colonising influences. An itinerant population meant no sustained or homogenous architectural forms, exacerbated by the lack of local building industries and the fact that all building materials have to be imported to the region. Despite this, the ACME team found other interesting and distinguishing elements to the region.

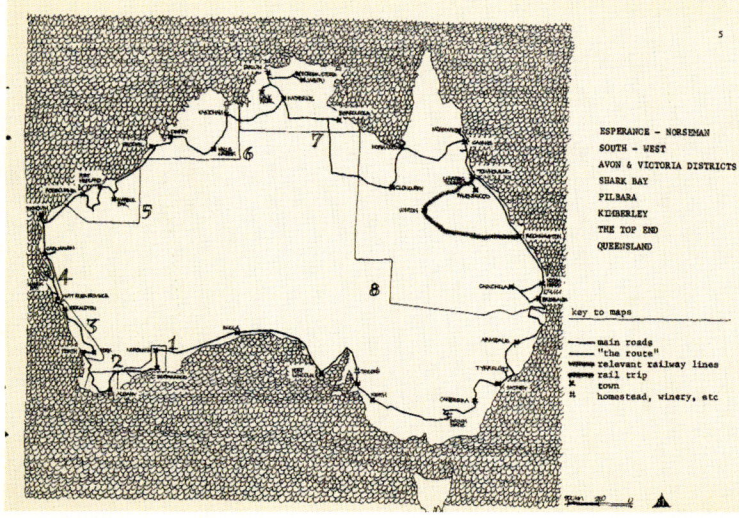

'wet' season daily cycle.
early morning
midday
afternoon

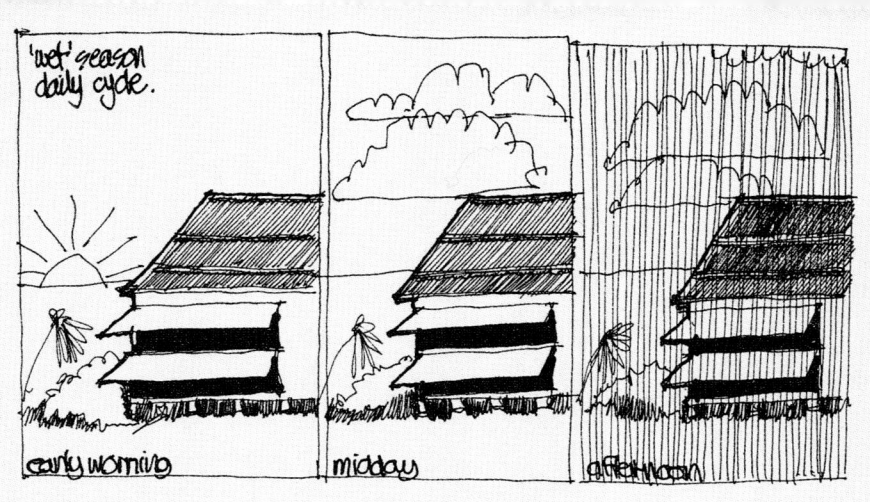

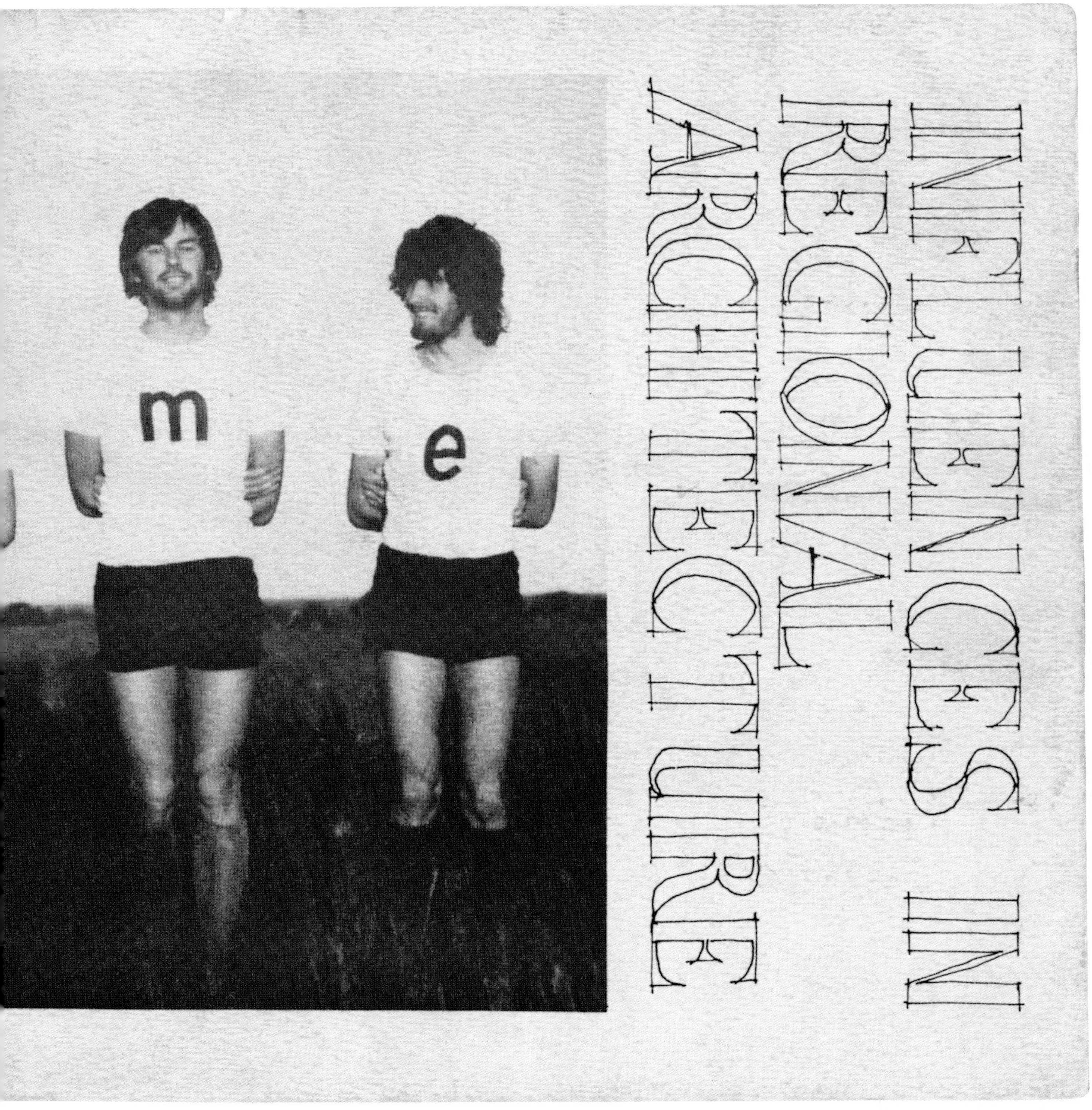
m
e
INFLUENCES IN
REGIONAL
ARCHITECTURE

During the 1870s, the first European shelters in the Top End were either tents or huts fabricated from locally found natural materials. By 1880, there was an influx of corrugated iron, that most Australian of prefabricated building materials. Corrugated iron sheet superseded Darwin porcellenite, the poor quality local stone which had been used to construct a small number of permanent structures of modest architectural pretensions. Parallel to this material shift was the importation to the region of Chinese workers from Singapore. The Chinese often used split or woven bamboo screening to enclose verandahs beneath corrugated iron roofs. The ACME team became fascinated with the local elevated houses that dated from the turn of the century. These were large rectangular houses raised on low stumps with a central core of rooms, and an encircling but enclosed verandah clad in vertical bamboo or timber slatting. Steep gabled or hipped roofs shed the rain during the monsoons. Slatted walls provided cross-ventilation and dappled sunlight. Very deep eaves and folding shade/shutter hopper window openings cast deep shadows and protected the wall of the house from the blistering sun. Of particular interest was the series of tropical houses designed in the late 1930s by government architect Beni Carr Glyn Burnett (1889-1955), especially a group of five elevated houses at Myilly Point in Darwin. Burnett, born in Mongolia, had arrived in Darwin in 1937. Educated in China and Edinburgh, he practised for many years in Tientsin (Tianjin), then Malaya before moving to Sydney and then being posted to Darwin. The houses at Myilly Point employed asbestos cement louvres with glass casement windows – fully screened walls could then be made infinitely adjustable according to the weather. Prefabricated army huts, asbestos cement louvres, flywire, the spaced palm trunks at Jabiru's hotel, and the idea of transportable architecture were all part of the rich and unsung architectural tradition of the Top End. All of these buildings captured the interest of ACME. They felt drawn to this sort of architecture, and also the sort of architectural response it implied – simple, no fuss and above all, the most direct way of dealing architecturally with the demands of extreme heat and extreme humidity.

The other aspect to Darwin was the devastating cyclone of Christmas Day 1974 when most of the city was literally blown away. When ACME visited Darwin, reconstruction was happening in earnest. Most of the new buildings, especially the houses, were of heavy mass construction with precast concrete units. It was an over-reaction to the cyclone and climatically inappropriate. Architects just didn't seem to be looking at their own local building traditions. They were ignoring an entire heritage of Top End architecture.

HOUSE BY BENI BURNETT AT MYILLY POINT C1940

sun
rain
humidity
cyclones

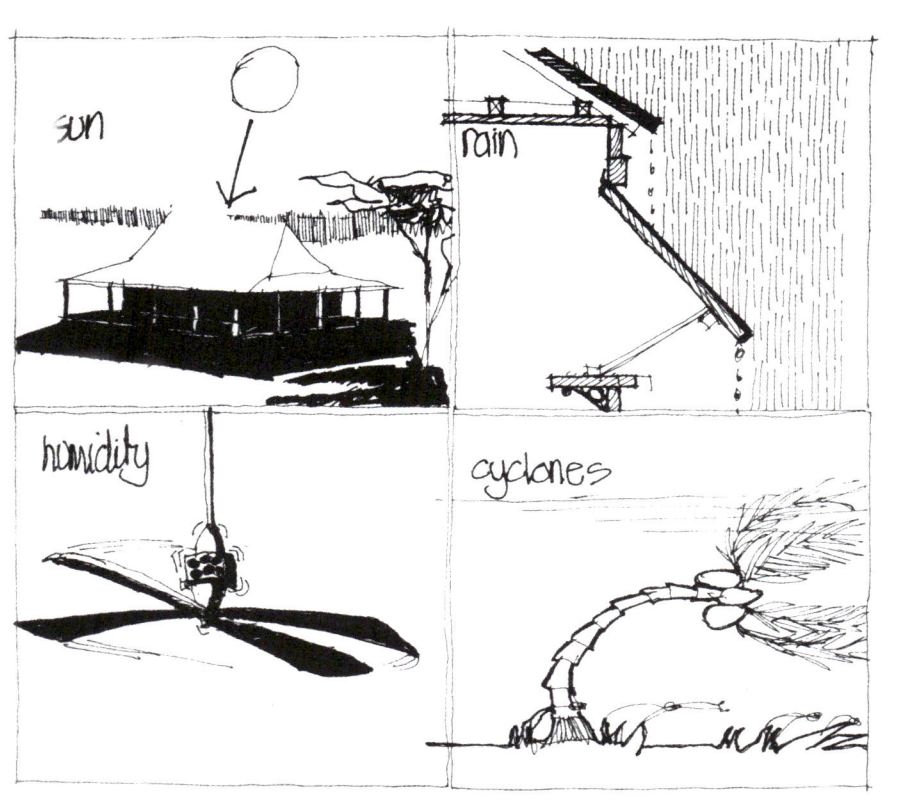

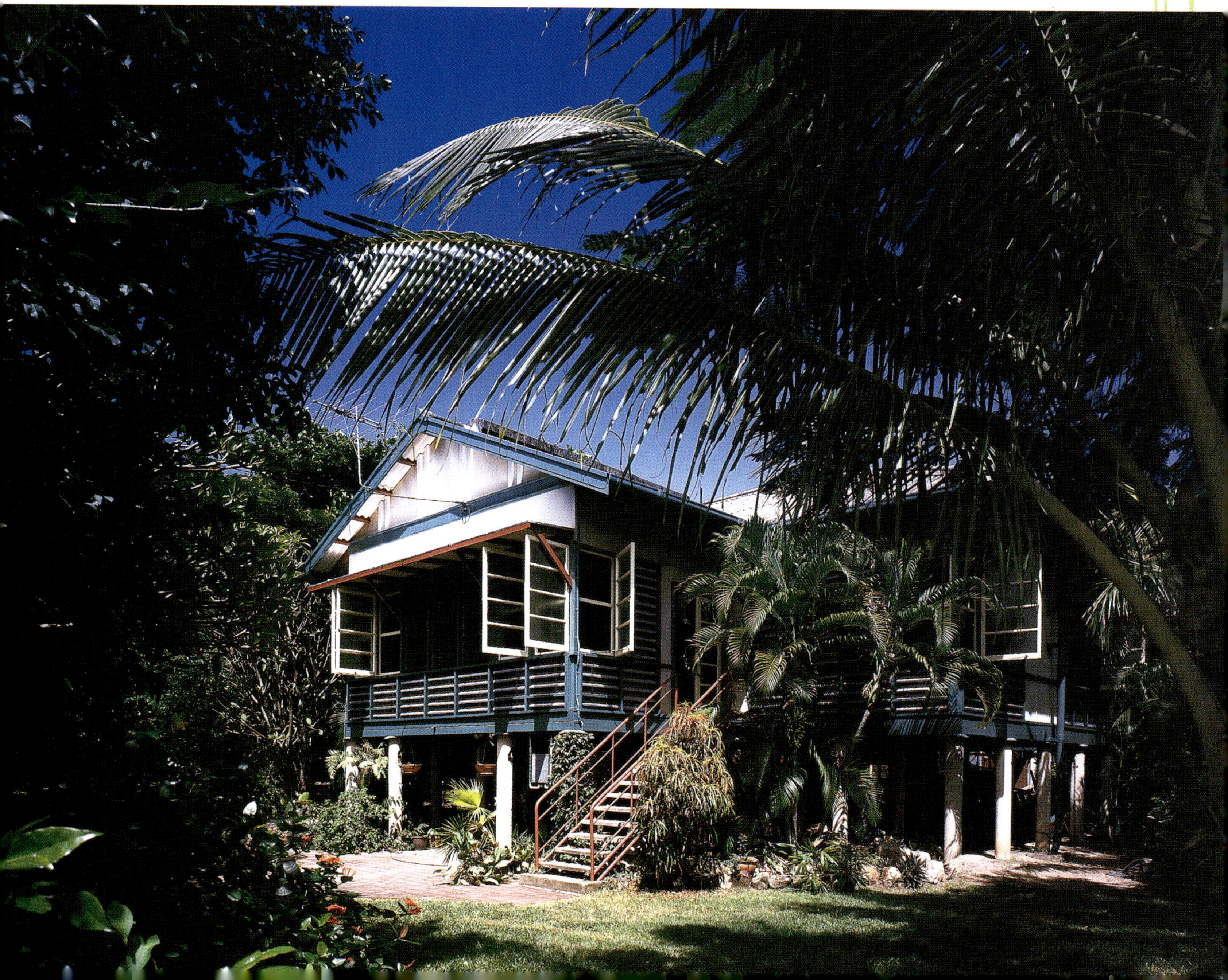

BUILT C1910, THIS SLATTED HOUSE WAS ORIGINALLY HOME TO AN EXECUTIVE PUBLIC SERVANT. RELOCATED SEVERAL TIMES, THIS HOUSE BREATHES, ITS EXTERNAL WALLS OFFER SHADE AND AIR, AND ITS ROOF SPACE IS OPEN ABOVE A CENTRAL CORE OF ROOMS.

The Move to Darwin

Back in Adelaide, by the end of 1978, there was little or no work for young graduate architects. Harris worked for no money in the recession-hit city for a dismal ten months. Welke left Adelaide and went back home to Esperance, WA to work on his parents' grazing property. They however convinced him to continue with architecture and after a brief stint in Perth, Welke returned to Adelaide to work for Russell & Yelland before shifting to Darwin, bonded to the Department of Transport and Works. He lasted six weeks. He then got a job with Vin Kenneally & Associates. Darwin, after the devastation of Cyclone Tracy, was experiencing something of a minor building boom, and nine out of sixty students in their year at university were working there. Four months later, on Welke's encouragement, Harris travelled north to Darwin and he too got a job in the Kenneally office.

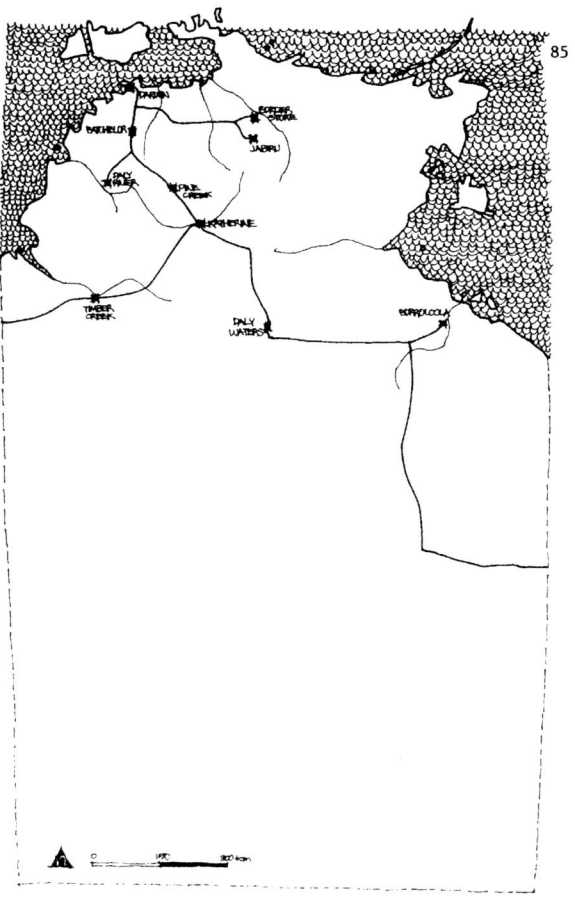

In mid-1980, Harris and Welke won a $2000 history grant to study housing in the Top End. Local historian Alwyn Horne had already begun to document prototypical government housing, but Harris and Welke wanted to collect together all the information on tropical housing as a way of determining a more generalised approach to housing design in the tropics. The outcome was that Harris and Welke set up a shopfront office in a lane off the Smith Street Mall to complete the study. Harris left the Kenneally office in September 1980. Welke did the same in December. Harris had taken the Coleman House job with him and it effectively became their first joint architectural commission. On February 11 1981, the office officially opened. To celebrate, they printed t-shirts with the new office name - Troppo Architects - plastered across their fronts. Troppo had begun.

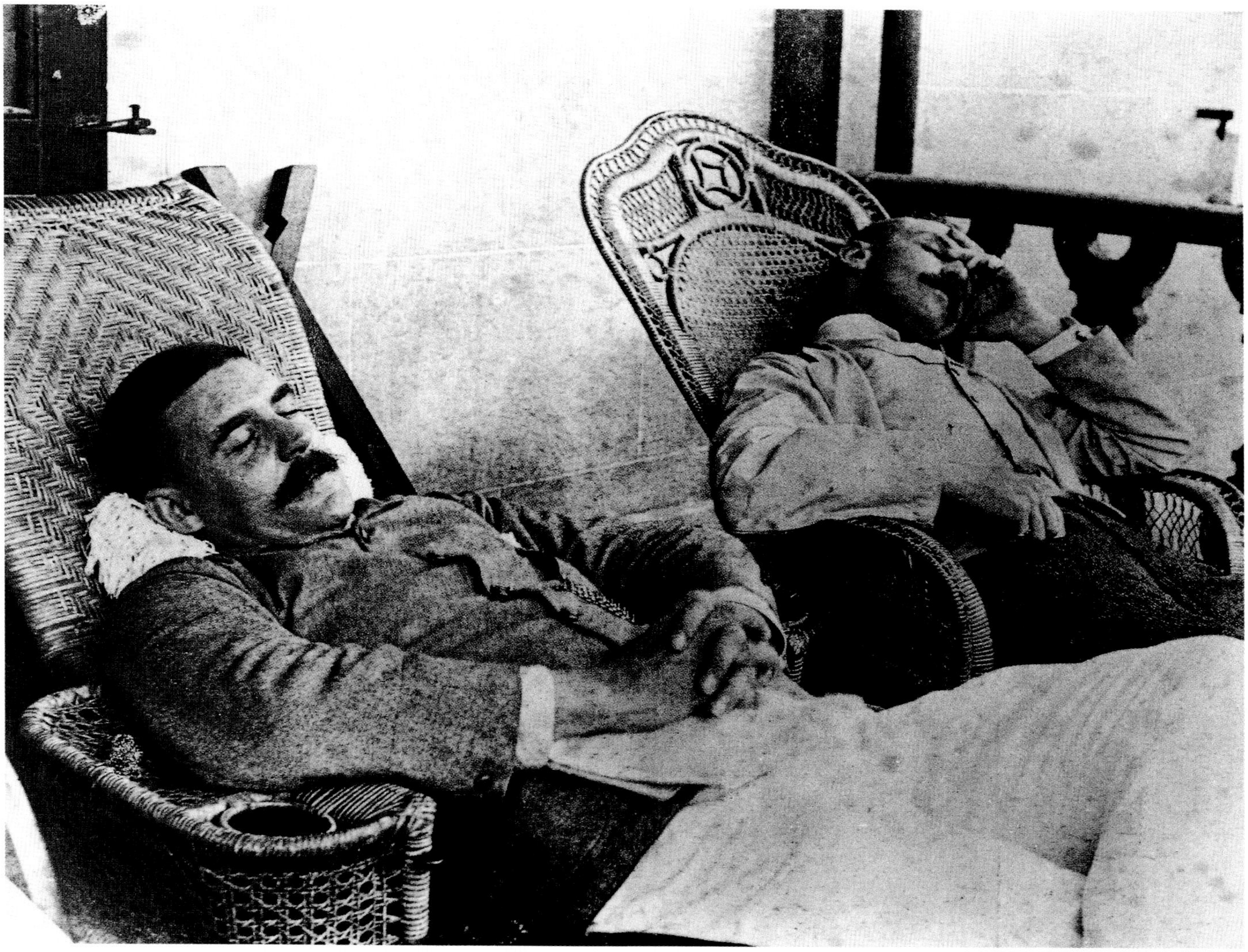

CHAPTER TWO
TROPPOVILLE

Establishment in Darwin

The Troppo office soon gained a reputation in Darwin - not immediately for architecture though. Two young blokes from Adelaide had joined a small town which had no organised recreational haunts. There was a social network of parties, drinking beer, and playing with the local cricket and football teams. Little work was to be had as yet, even though Troppo were publishing regular articles in the local newspapers about better tropical house design. Troppo shared its shopfront office with freelance playwright, Simon Hopkinson (eventual scriptwriter of 'Bananas-in-Pyjamas') who was Director of the Darwin Theatre Group. Together they designed sets for a number of plays which were performed in one of the WWII gun emplacements at the East Point Reserve. Troppo learnt quickly that they were part of a community where great respect was due to those locally born, and where the economy was to a large degree propped up by forces from elsewhere which could disappear at any time. Darwin was also a place where many people had come to escape, even disappear. For almost two years Troppo didn't have a car, they had bicycles and two PMG scooters. It was a shoestring operation. They shared houses and charged their architectural work out at an hourly rate.

FITZROY CROSSING HOTEL

Naturally tropical

- This house in Schultze Street, Larrakeyah, is an example of the now rapidly-fading tropical architecture of pre-war Darwin.

Living in cool houses

- This house featuring bamboo shutters around a wide verandah was probably built in Darwin in the 1930s. It no longer exists.

- A railway cottage still standing at Pine Creek -- a core of three rooms are surrounded by wide verandahs. The iron cladding in this design maximises cool living.

THE transplanted southern homes that squat solidly in hot rows, typical of Darwin's suburban populations, are about to find rivals in styles harking back to days of natural cooling.

Darwin used to have a distinct architectural stamp -- a unique style and character still seen in some houses in Larrakeyah, the RAAF Base and dotted through the older suburbs.

These houses are the remnants of pre-airconditioning and are designed to suit the climate, rather than fight it or block it out.

Troppo Architects, the brainchild of Adrian Welke and Phil Harris, is designing and building houses based on the principles of these old homes of the 1930s and 1940s.

They have studied plans still to be found in government departments and have come up with a basic blueprint modified to suit individual style.

"The buildings are responsive to the time and place," says Adrian.

They make the most of shade, breezeways, large verandahs and lightweight materials such as timber sheeting, colorbond and galvanised iron.

Shutters, canvas and shadecloth provide screening from the sun while allowing maximum airflow.

In the tropical house of the 1930s a core of three or four rooms was surrounded on all sides by wide verandahs, where most of the living was done.

They were designed to get wet, and people moved around the space catching the breeze and keeping dry.

"It's a lifestyle that perhaps not everyone would like," says Adrian.

"But it offers logical solutions to a climate that can be uncomfortable.

"But no-one is advocating these designs.

"Developers in the northern suburbs are building the only thing that people already understand, and they are not interested in offering anything more in accord with Darwin."

Troppo is.

Four of its designs are already being built, and the architects say they are flat out designing more.

One design is included in the experimental low-cost housing scheme sponsored by the government.

This core house will cost around $30,000 in a linear design that will trade on room for the homeowner's ideas in lattice, canvas awnings and other do-it-yourself additions.

They call it dynamic building -- a kit-style home that allows one's own personality to stamp a response to the climate.

Troppo's shopfront office off The Mall is planned to attract the passing crowd.

"It's important to take housing ideas to people and to suggest that Darwin's architectural heritage can be re-interpreted to suit people now," says Adrian.

by SUE WILLIAMS

- It's still standing at the RAAF Base ... an example of 1930s government housing that provided coolness through cross ventilation, shutters and wide eaves.

WHATEVER THE WEATHER, WE'VE GOT YOU COVERED.

Buy Now and Save!

If you're quick, you'll be able to buy total protection for your home now -- and keep it cool all year round. Take advantage of this saving now!

Phone us now for an obligation free measure and quote.

DABSCO PTY LTD
1546 WINNELLIE RD, WINNELLIE
PHONE 84 3111

TROPPO HOUSE, COCONUT GROVE

TROPPO HOUSE, COCONUT GROVE 1981

In April 1981, Troppo bought and moved into an old 1920s railways house. They had it relocated to a site in Coconut Grove, Darwin. Harris, his partner Annie and their new baby Patrick took one half while Welke and his girlfriend Fran Bonney (they married later that year) took the other half. They added four wings: two sleeping porch/verandahs; a kitchen living wing and a creek porch. Their purchase typified a way of procuring a house in the Top End. At that time in the Northern Territory, $50,000 could be borrowed from the Territory government at 4% interest. This invariably meant a package of $20,000 for land and $30,000 for the house. Troppo paid $450 for the house and it cost $7,000 to move. The rest of the money they used to make the additions. There were many lessons to be learnt from such a pragmatic process, with no cause to be precious about building anew or owning something pre-loved. Architecturally there were also new challenges. Troppo began to understand many things about designing for such a place. They realised the importance of single-skin construction, having no kickboards to remove any chance for cockroaches to live and hide, of elevating all the cupboards, of minimising all roof spaces, of minimising rafter dimensions for economy, and of trying to avoid ceiling linings which could become home to dead animals, bats and birds. They also experimented with louvres, open porches, and splitting floor levels to introduce gaps for breezes to pass through the house. They also began to appreciate the house as being an organic entity, something which could grow, something which in the tropics implied almost no physical boundaries to impede airflow.

'Punkahs and Pith Helmets'

At the end of 1981, Troppo completed their study of Top End housing. The results of the study were self-published in June 1982 as *Punkahs and Pith Helmets: good principles of tropical house design*. It was more detailed than their earlier survey. They had worked hard to re-examine in more detail the buildings that they had seen two years earlier. Troppo included studies of aboriginal shelters and noted the importance of sun and wind in determining their shape and structure. They also noted the pre-European presence of Macassan fishermen from the South Celebes who came annually to the Top End in search of trepang (sea slug) and who built temporary stilted shelters with steeply pitched roofs. The study gave them another opportunity to examine more closely the plans of Burnett's houses at Myilly Point. They also discussed the importance of the stilted Queensland house as an influence on both government and private housing; the pervasive presence, due to World War II, of prefabricated corrugated iron Sidney Williams huts; and the pragmatic planning and construction of single men's quarters in mining settlements during the 1950s and 1960s.

THE REGIONAL CONTEXT
.1. a history of top end housing

SHELTER PRE-EXISTENT TO WHITE SETTLEMENT

Aboriginal shelter, apart from sacred cave sites, has been created only out of climatic response, and is of less importance in site usage than the campfire and outdoor living. Because of this, aboriginal shelter exhibits a very simple translation of climatic priciples.

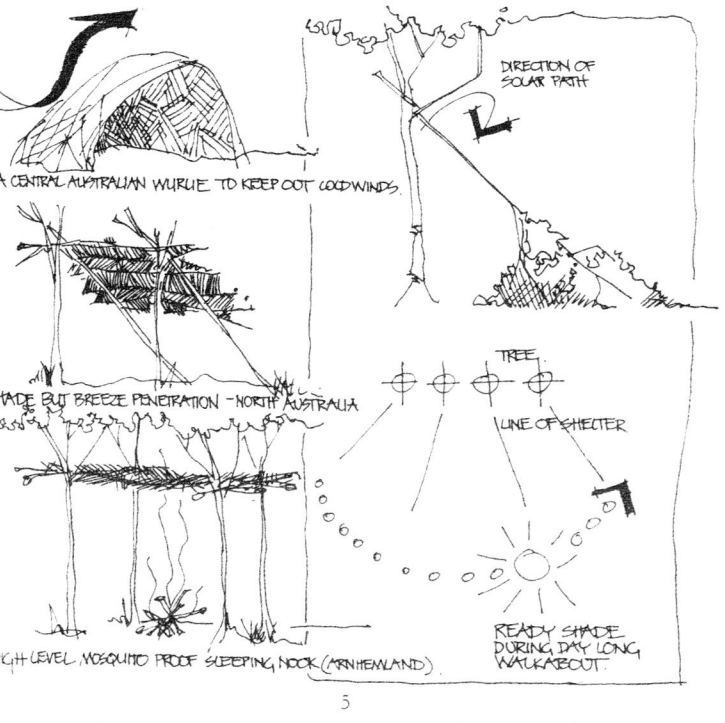

Principles of Tropical House Design

The major part of *Punkahs and Pith Helmets* was devoted to principles of tropical house design developed through their empirical research, photographs, and drawing studies of 'tinnies', corrugated iron houses and sheds, and elevated houses. This research was complemented by Troppo's readings of theorists like Habraken, Sommer, Chermayeff, Alexander, Papanek, and Rapaport where issues of spatial hierarchy, shelter, adaptability, access, and physical comfort were crucial elements of a more humane and unpretentious architecture.

Troppo proposed four fundamental architectural principles: 1) the promotion of cooling breezes; 2) ventilation by convection; 3) reducing radiation of heat, and; 4) the sheltering of walls and openings. They illustrated each of the principles in more detail with the engaging freehand sketches which have characterised their work ever since. Issues such as minimising any internal barriers to cross ventilation within the house; incorporating supplementary means for ventilation such as rotary vents or windtraps; the various techniques for venting roof spaces; the need to minimise insulation to encourage thermal advantage from any small variation in temperature that might occur; and the protection of walls from sun in the Dry and water in the Wet were all covered. The structural concerns of bracing, framing and holding down for cyclones, the architectural effect which could achieved by the frame exemplified by the post and beam, and the infill panels of the traditional Japanese timber house were also described. These principles in effect laid down ground rules for the development of a regionally and climatically responsible architecture. They were to inform the next twenty years of Troppo's design practice.

A. PROMOTION OF COOLING BREEZES.

Given a cooling path for breezes, apparent reduction in temperature can be readily achieved. Firstly this occurs through replacement of air which would otherwise increase toward outside temperatures and secondly through increasing the rate of heat loss from the body by the motion of air past the skin.

To maximise the effect of cooling breezes it is necessary to look at the house in 4 ways:

i   in its presentation toward prevailing breezes

ii   in interaction with the outdoors

iii in treatment of potential barriers to the path of breezes through the house

iv   in incorporating supplementary means of promoting air movement

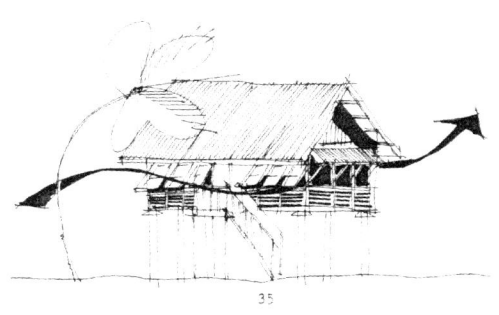

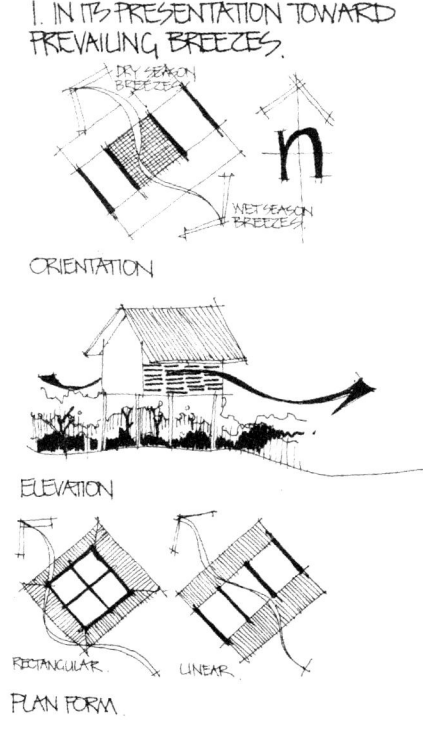

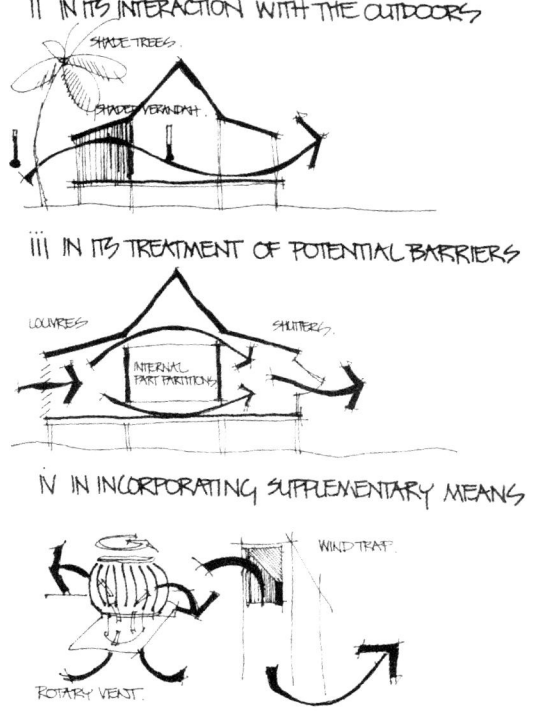

GREEN CAN 1980

The 'Green Can'

To demonstrate the tropical house design principles, Troppo included in *Punkahs and Pith Helmets* a design which they had completed in September 1981 as part of a Low Cost House Competition. A village of eleven competition winners was built in an outer Darwin suburb in May 1982. Troppo's design was humorously nicknamed 'The Green Can' after the signature green Victoria Bitter beer can. Built to cost no more than $34,000, the Green Can was a challenge to all contemporary house designs then being constructed in Darwin. It was deliberately provocative, and many in the local community clearly thought so too. Talk-back radio in Darwin found harsh critics of Troppo's new tropically responsive house.

The Green Can follows all the Troppo principles. Oriented to catch the wet season breezes from the north west, and the dry season breezes from the south-east, the house is distinguished by its steep opposing skillion roofs (35 degrees optimum design pitch for windloading). The higher one is over the bedroom and living spaces (called rooms 1, 2, 3 on plan to denote flexibility), and the lower one is over the bathroom, laundry and kitchen. Each of these zones is separated by a 'colonnade', an open passage which widened in the centre of the house to become a breezeway. It is essentially a roofed outdoor room which enables cross-ventilation and spatial expansion across the entire width of the house. Open, adaptable, with lightweight exposed steel framed construction, corrugated iron cladding and timber lattice infills, the Green Can proved that Troppo's formula for architectural research had substantial merit. It simply required Darwin people to accept that it had a possible future as an appropriate tropical house.

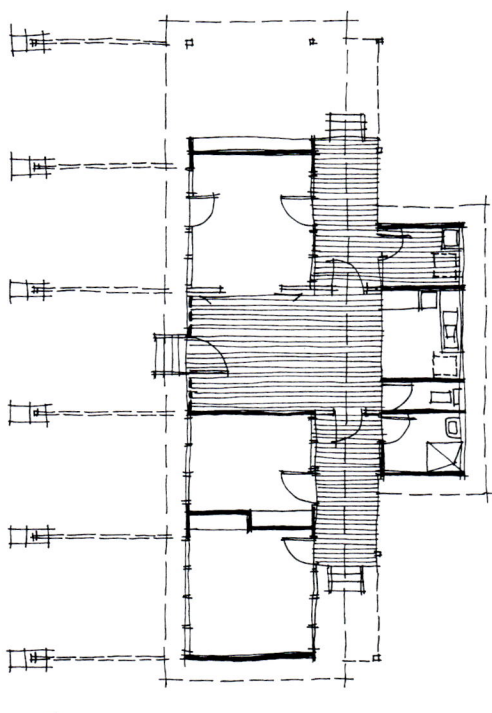

GREEN CAN 1981 PLAN

GREEN CAN 1981 SECTION

"WORTH HOUSE (ELEVATED GREEN CAN)" COCONUT GROVE 1981

ELEVATED GREEN CAN INTERNAL

DRAPER HOUSE 1982 ELEVATION

Early Houses

The Green Can gave rise to several individual house commissions. Most demanded to be variations on the original type, and this was Troppo's ideal. The Green Can was not meant to be understood as a fixed entity, but as the fluid basis for an entire range of house types. Six Green Can variations were built and many were elevated models of the type. The Gow/Royale House, Howard Springs (1982) and Draper House, Leanyer (1982) both followed closely the original Green Can principles, while the Worth House (1981) at Coconut Grove was an elevated Green Can.

ELEVATED GREEN CAN (UNDER CONSTRUCTION)

DRAPER HOUSE 1982 ELEVATION

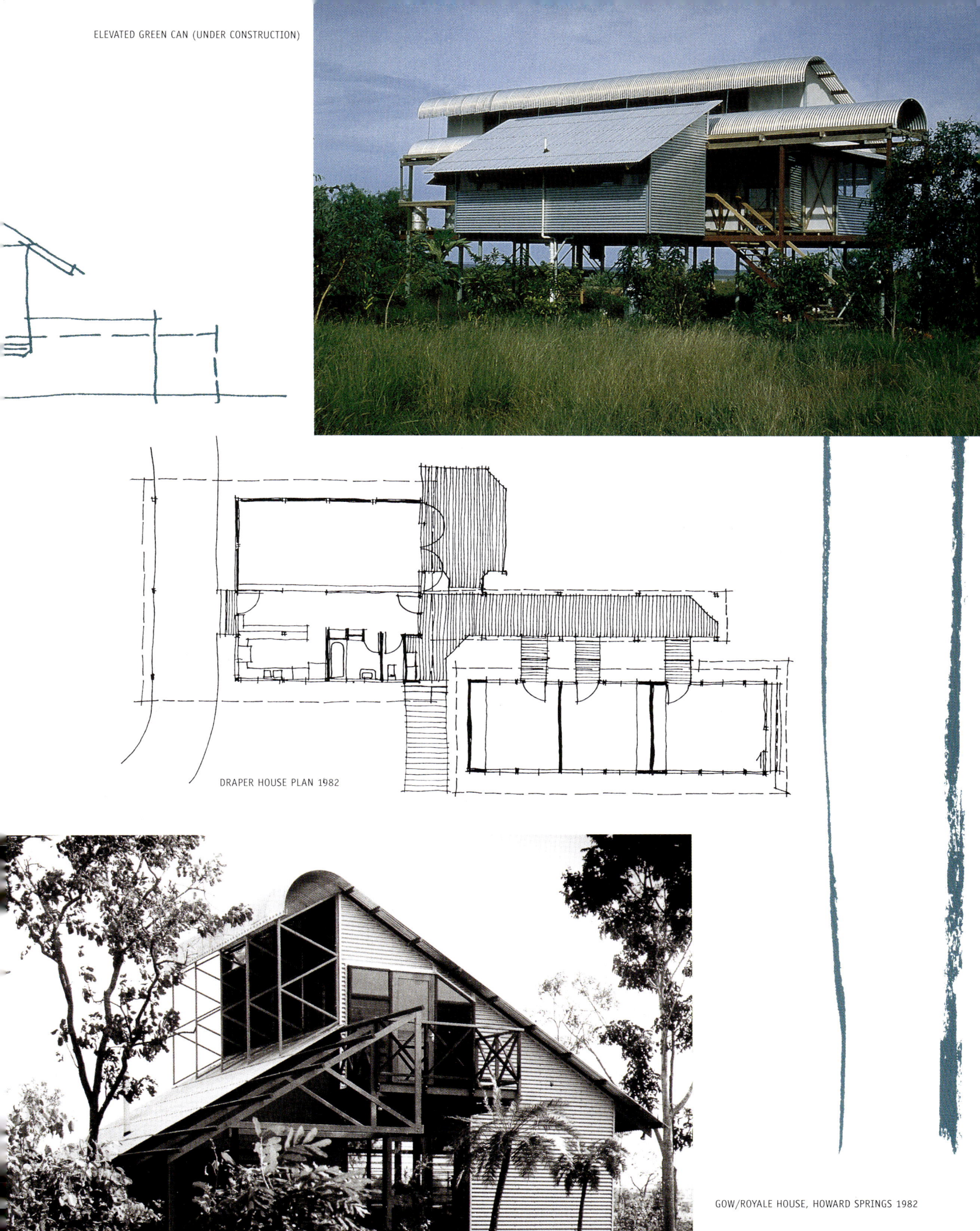

ELEVATED GREEN CAN (UNDER CONSTRUCTION)

DRAPER HOUSE PLAN 1982

GOW/ROYALE HOUSE, HOWARD SPRINGS 1982

LAWLER HOUSE, COCONUT GROVE 1980

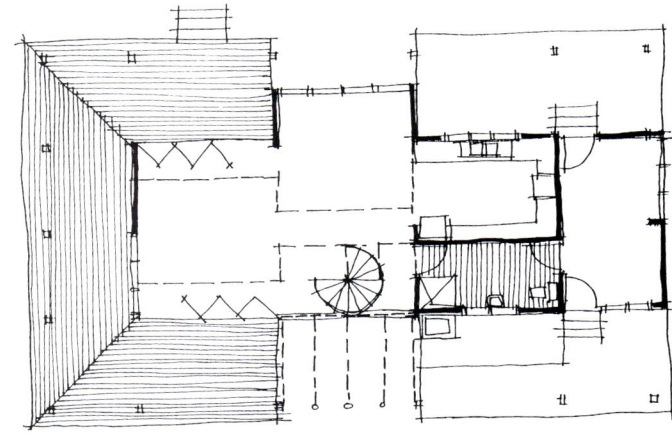

BUTCHER HOUSE, VIRGINIA 1982 PLAN

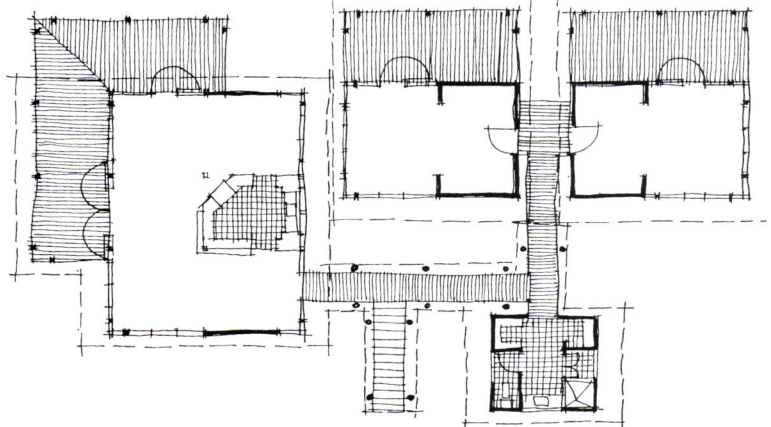

LAWLER HOUSE, COCONUT GROVE 1980 PLAN

By 1983, five Troppo houses had been built in Coconut Grove and the suburb had been nicknamed Troppoville by the locals. Houses like the Coleman House (1980), designed earlier but finished later than the Green Can, followed more traditional notions of the stilted Queensland house. With its broad timber decks doubling as living spaces, the Coleman House included an extensive use of louvres, inner walls reaching only half-way to the 6 metre ceiling, and a similar barrel vaulted roof venting system to the Green Can. The Kaiplinger House (1983) comprised two elevated pavilions (one living, one sleeping) with an entry point defining the split between either zone. The dark stained timbers, bookshelves held up high and a sense of total openness amidst the jungle-like surrounding garden exemplify Troppo's versatility in adapting their principles to individual siting and client requirements as well as to prototypical design.

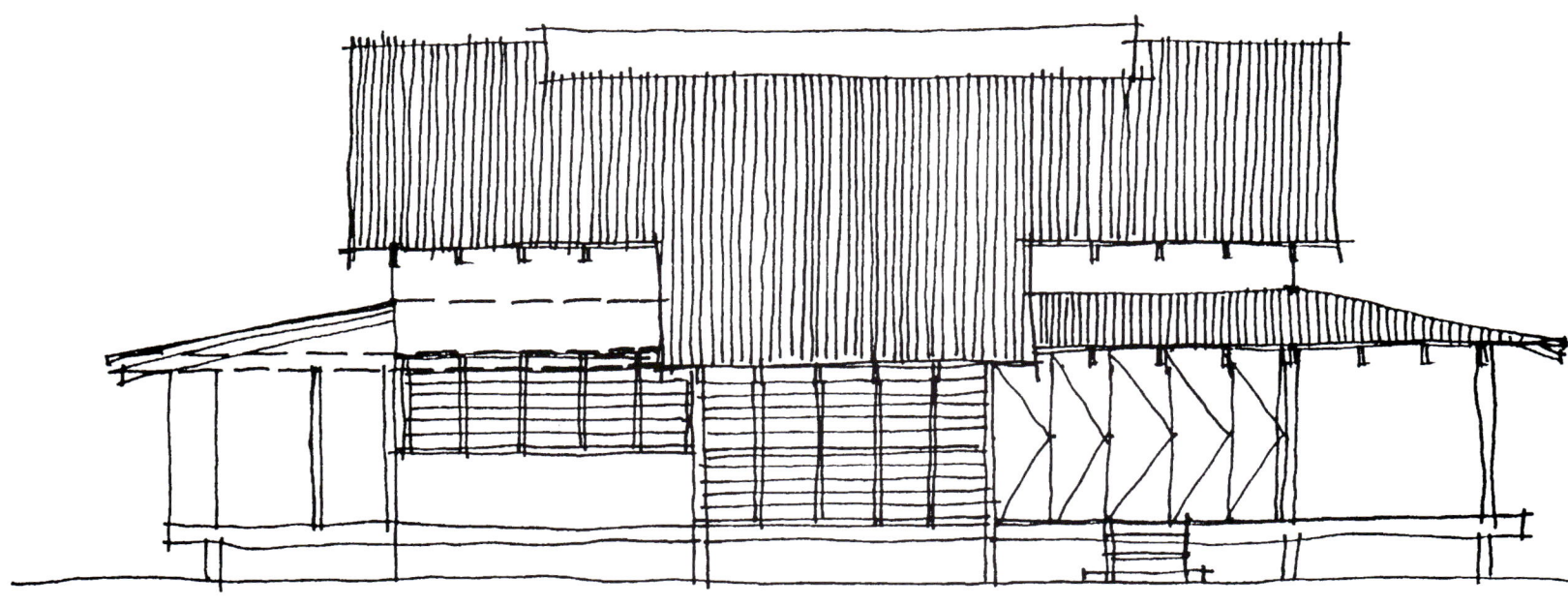

BUTCHER HOUSE, VIRGINIA 1982 ELEVATION

LAWLER HOUSE, COCONUT GROVE (UNDER CONSTRUCTION) 1980

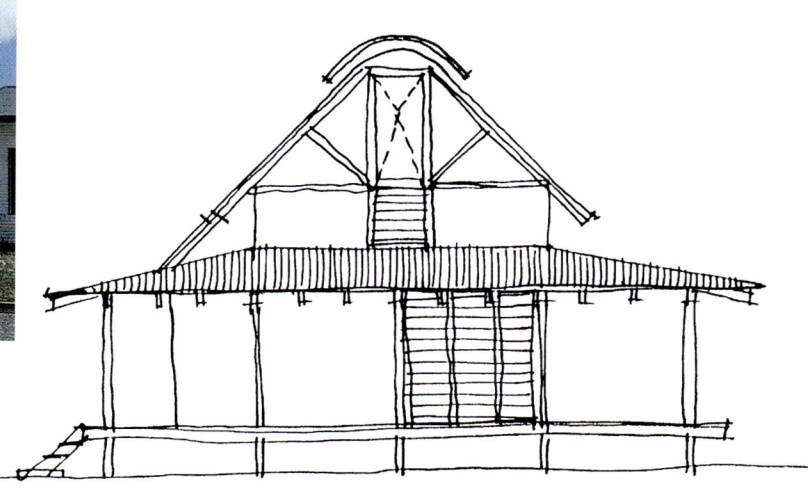

BUTCHER HOUSE, VIRGINIA 1982 ELEVATION

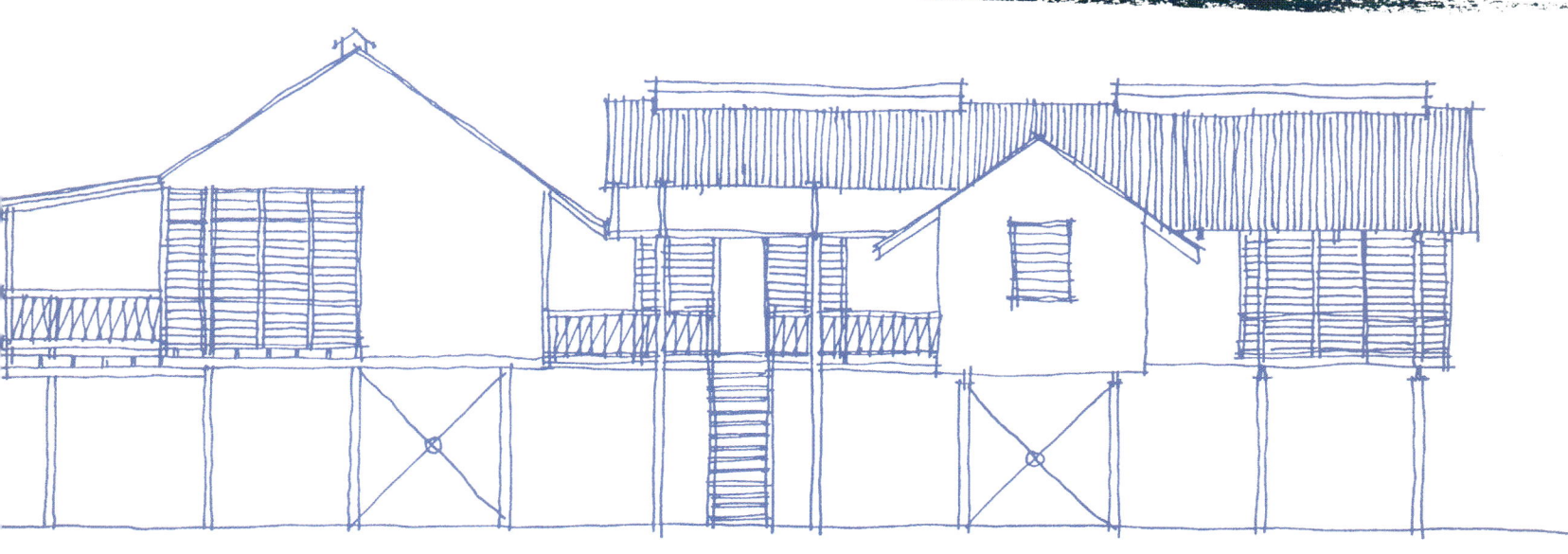

LAWLER HOUSE, COCONUT GROVE 1980 ELEVATION

KAIPLINGER HOUSE, COCONUT GROVE 1983

KAIPLINGER HOUSE, COCONUT GROVE 1983

KAIPLINGER HOUSE, COCONUT GROVE 1983 ELEVATION

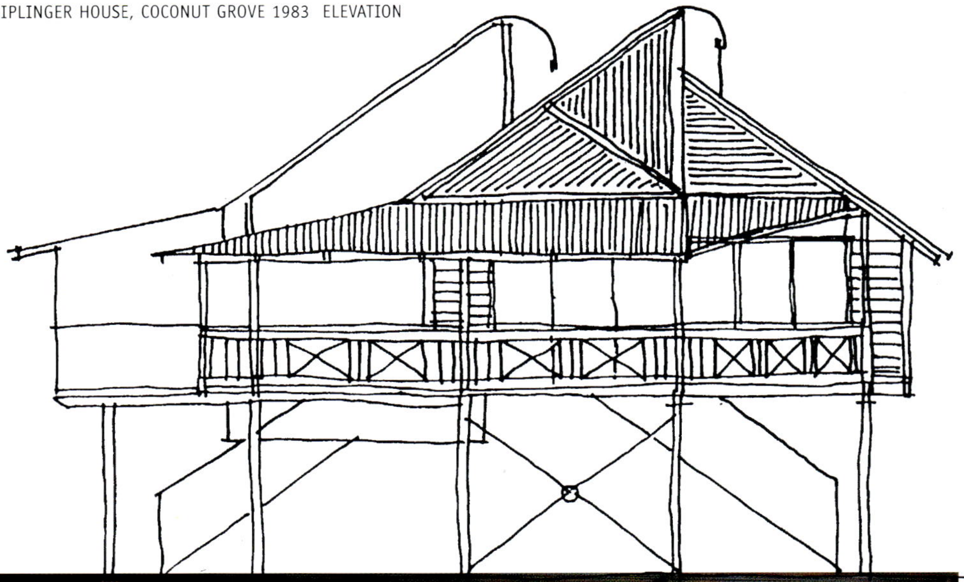

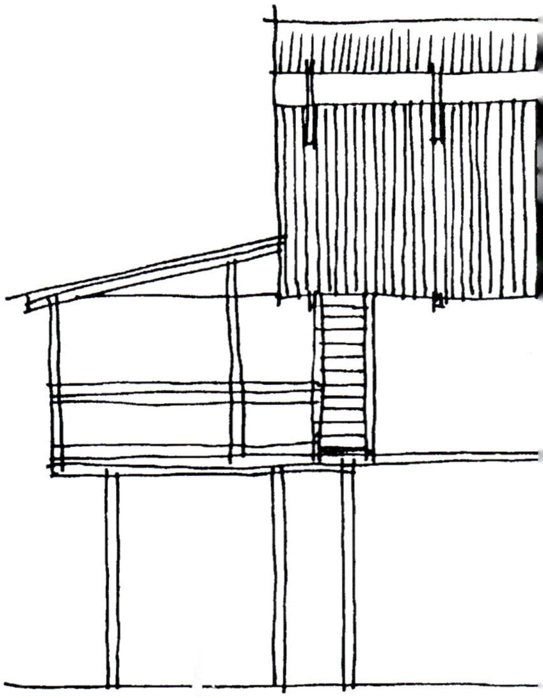

KAIPLINGER HOUSE, COCONUT GROVE 1983 ELEVATION

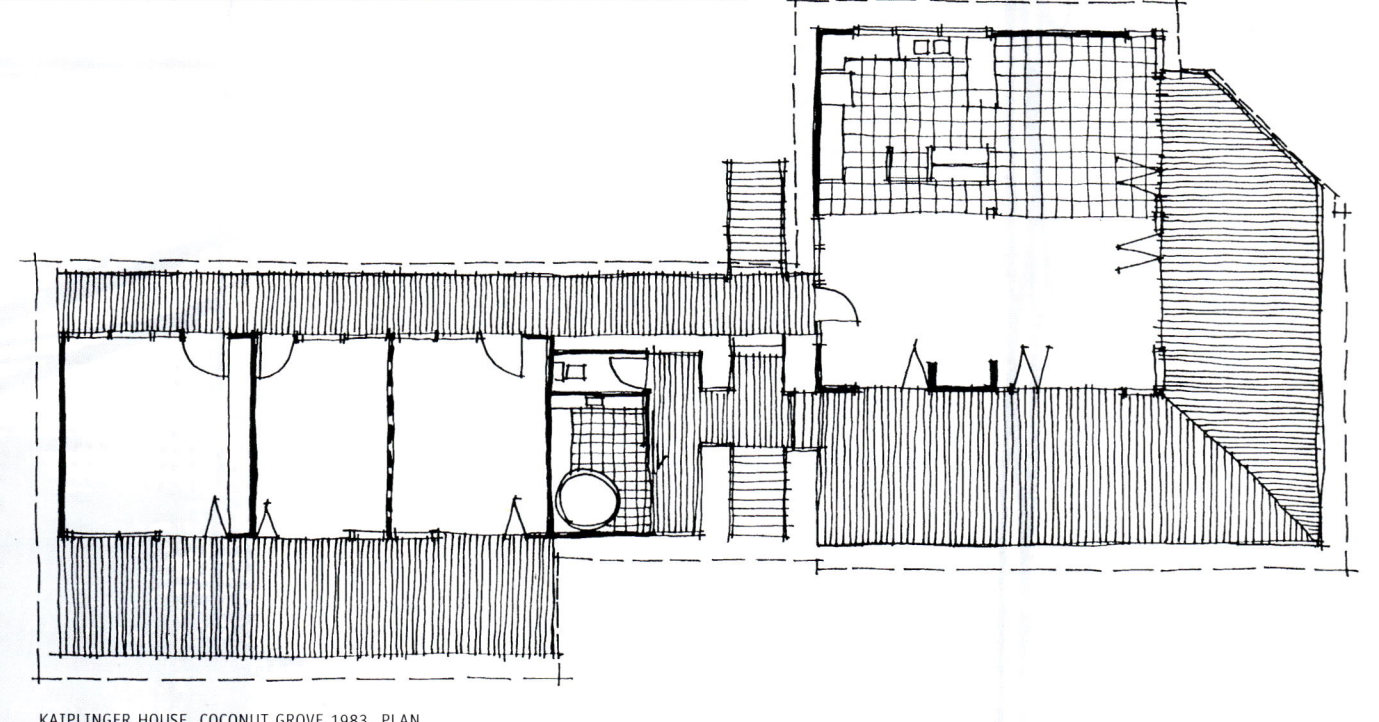

KAIPLINGER HOUSE, COCONUT GROVE 1983 PLAN

RHODES HOUSE, NORTHLAKES 1986

PITT HOUSE, PALMERSTON 1983 ELEVATION

More single family houses followed and Troppo determined each design according to stringent budget controls with the minimising of materials and waste. A discipline of lean informality came through working to budget and making do. A corresponding aesthetic developed with detail inspiration sometimes drawn from old tropical architectures which Troppo knew well. Although reassured by the emerging cohort of architects elsewhere in Australia like Russell Hall, Rex Addison, Gabriel Poole and Glenn Murcutt whose work showed similar tendencies to their own, Troppo prided themselves on being different. Their architecture was tougher because it had to be loose, adaptable, and effortlessly buildable, though still able to achieve the distinction of being an architecturally informed shelter. The Rossetto House, Northlakes (1985), brought the entry veranda inside behind a screened formal facade. From the street, this house appeared as a slatted prism. By contrast, the Pitt House, Palmerston, (1983) with its steep pitched roof and exposed diagonal timber struts was a series of pavilions connected by walkways. For each client, Troppo's response to type could vary - the kit of parts could be specifically melded. By 1990, a poster of all of Troppo's houses indicated mock parity with the permutations of Palladian villas and Florentine palazzi. This also signified that their architecture defined not just the development of type, but an architecture of place.

ROSSETTO HOUSE, NORTHLAKES 1985

PITT HOUSE, PALMERSTON 1983

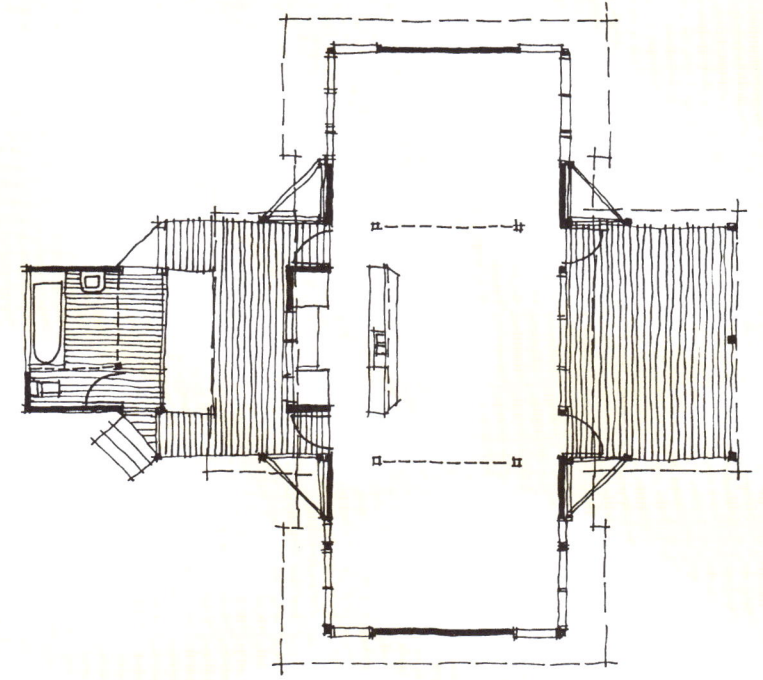

PITT HOUSE, PALMERSTON 1983 PLAN

GEROVICH HOUSE, PARAP 1984

ST ANDREW'S LUTHERAN SCHOOL (DUNNY), LEANYER 1982

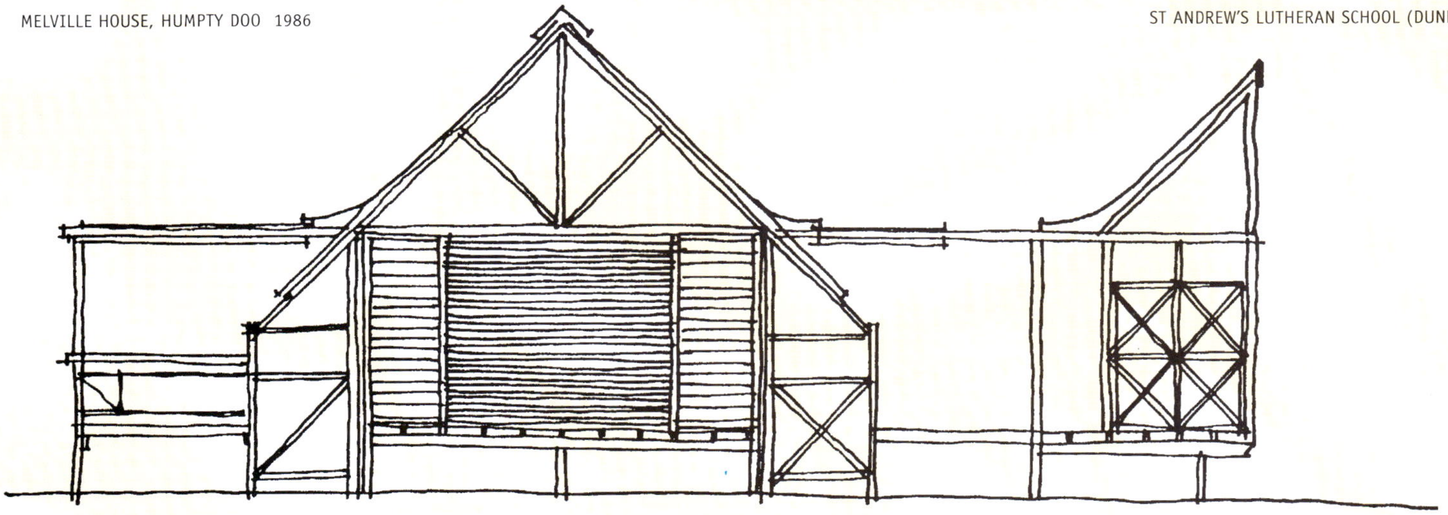

MELVILLE HOUSE, HUMPTY DOO 1986

PITT HOUSE, PALMERSTON 1983 ELEVATION

HAZELDINE HOUSE, NORTHLAKES 1989: AN EARLY EXAMPLE OF HOUSE AS COMPOUND, THE HAZELDINE HOUSE COMPRISES THREE SEPARATE STRUCTURES: LIVING/MASTER BEDROOM; BATH-HOUSE; AND CHILDRENS' WING. IN THE MAIN LIVING SPACE, A TREE TRUNK FORMS A DRAMATIC STRUCTURAL PROP.

UPPER LEVEL PLANS

ELEVATION

ELEVATION

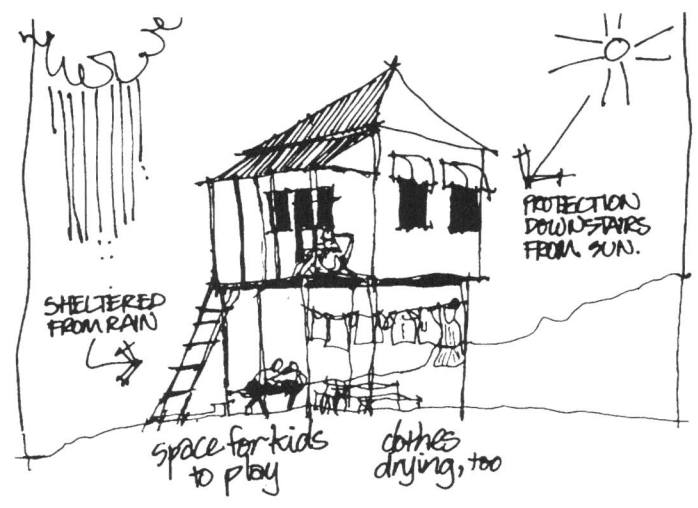

During the early 1980s, Troppo worked hard to publicise themselves and the benefits of responsible tropical house design. They continued their weekly newspaper articles. They would be published wearing pith helmets with beer cans in hand. Theirs was a special form of advertising — always done with irony, tongue in cheek and the clear message that a good time was to be had. This was not to undermine the seriousness of their intent. Troppo's involvement with Darwin's architectural heritage intensified in the mid-1980s with moves to raze the Burnett houses at Myilly Point to build a second casino. Troppo lobbied hard, rallied, even against the National Trust and the local chapter of the RAIA so that the houses could remain on their sites and not be moved. Their success meant the creation of a heritage precinct at Myilly Point, and a resounding point being driven home about the continuing relevance of Darwin's 1930s tropical housing to future generations.

In 1983, Troppo moved from its shopfront office to a 1920s slatted house, a former public servant's house and this became the office for the next five years. In 1988, they moved again to occupy another government house, a G-type designed by Burnett for an executive public servant. As with the slatted house, this house in Lindsay Street was to become part of Troppo's empirical experience as to the benefits and workability of previous designs for the tropics. Troppo were working directly within typologies which they sought to understand, learn from, and distill design principles.

TROPPO ARCHITECTS
TROPICAL HOUSES OF NORTHERN AUSTRALIA
1980-1990

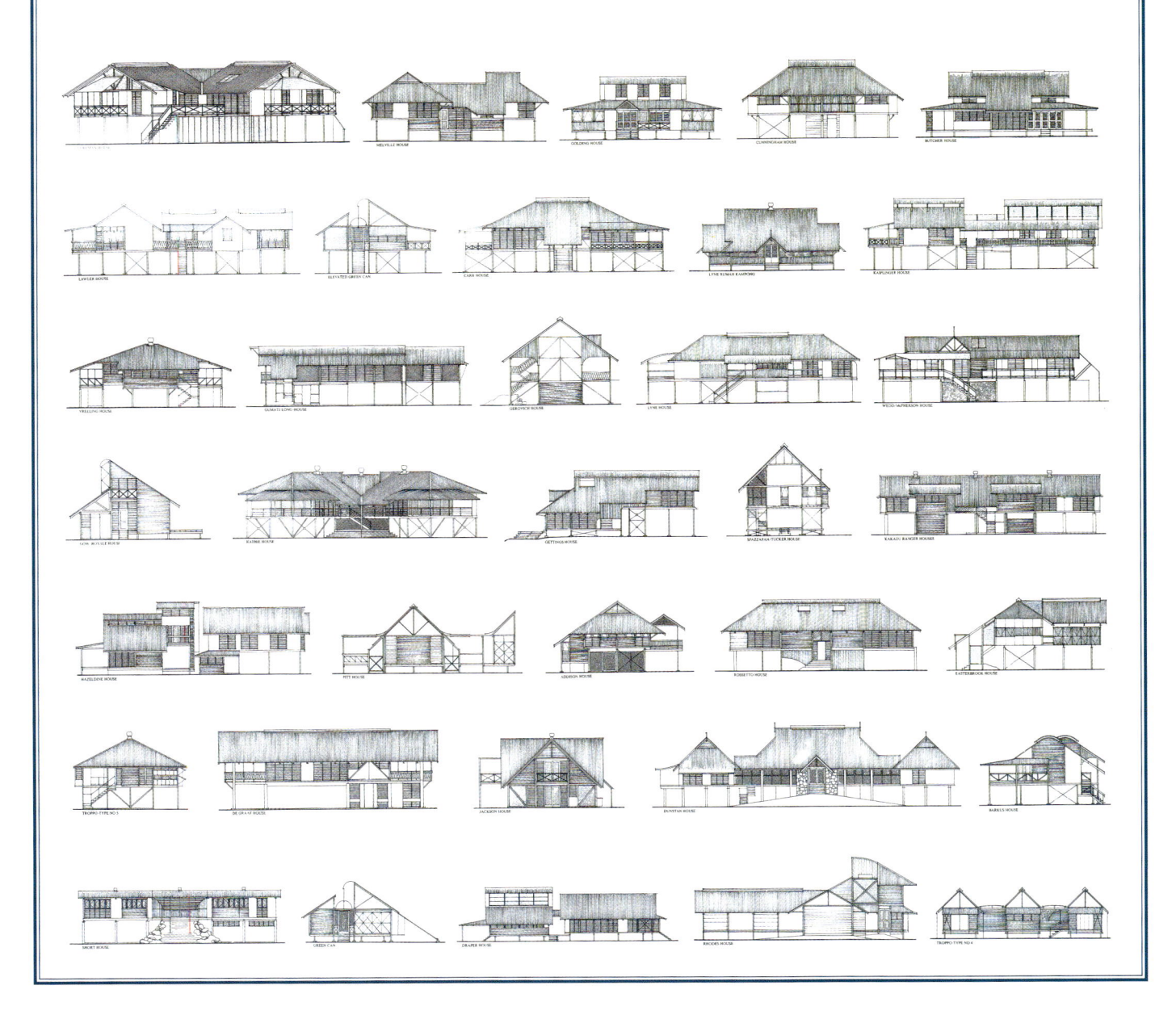

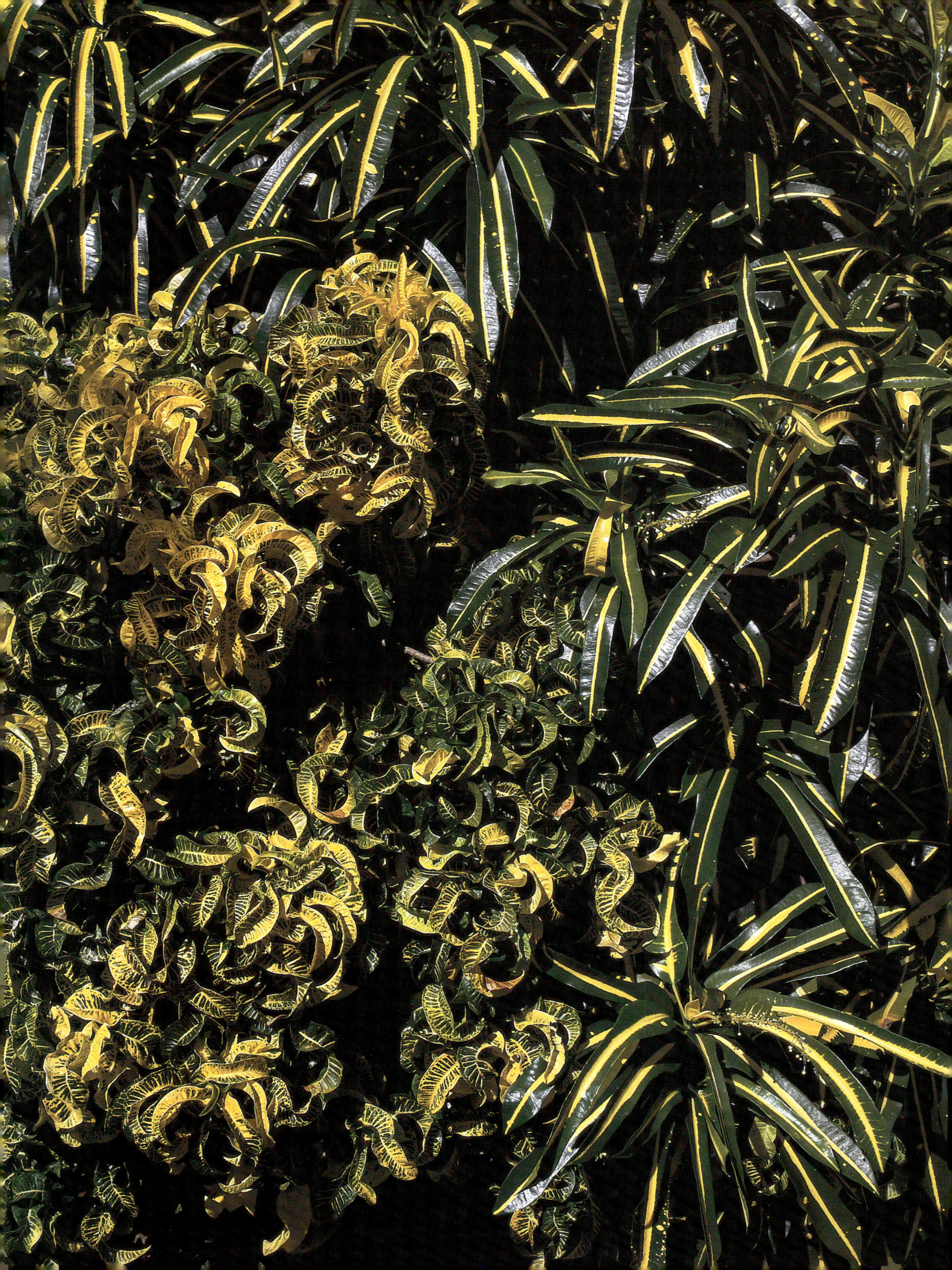

CHAPTER THREE
NEW CHALLENGES

From 1989, new sorts of commissions began to come Troppo's way - and from very different sources - the Federal Government's Defence Housing Authority, and the aboriginal communities of remote Arnhem Land. There was also another important house design competition win for Troppo - the aptly named Tropical House (also known as Troppo Type 5). Important to this house was the notion of collaboration. The Troppo office has always had a number of people passing through, collaborating at various times. In Darwin's past, many architects, including its most notable temporary resident architects - JG Knight and Beni Burnett, stayed briefly, made a contribution and then passed on. With Troppo, Danny Wong of Speargrass Architects, Sue Harper, and Glenn Murcutt would also be part of this process. The relaxed and inviting environment of the Troppo office made collaboration easy, necessary and it gave the office periodic injections of different sorts of vitality.

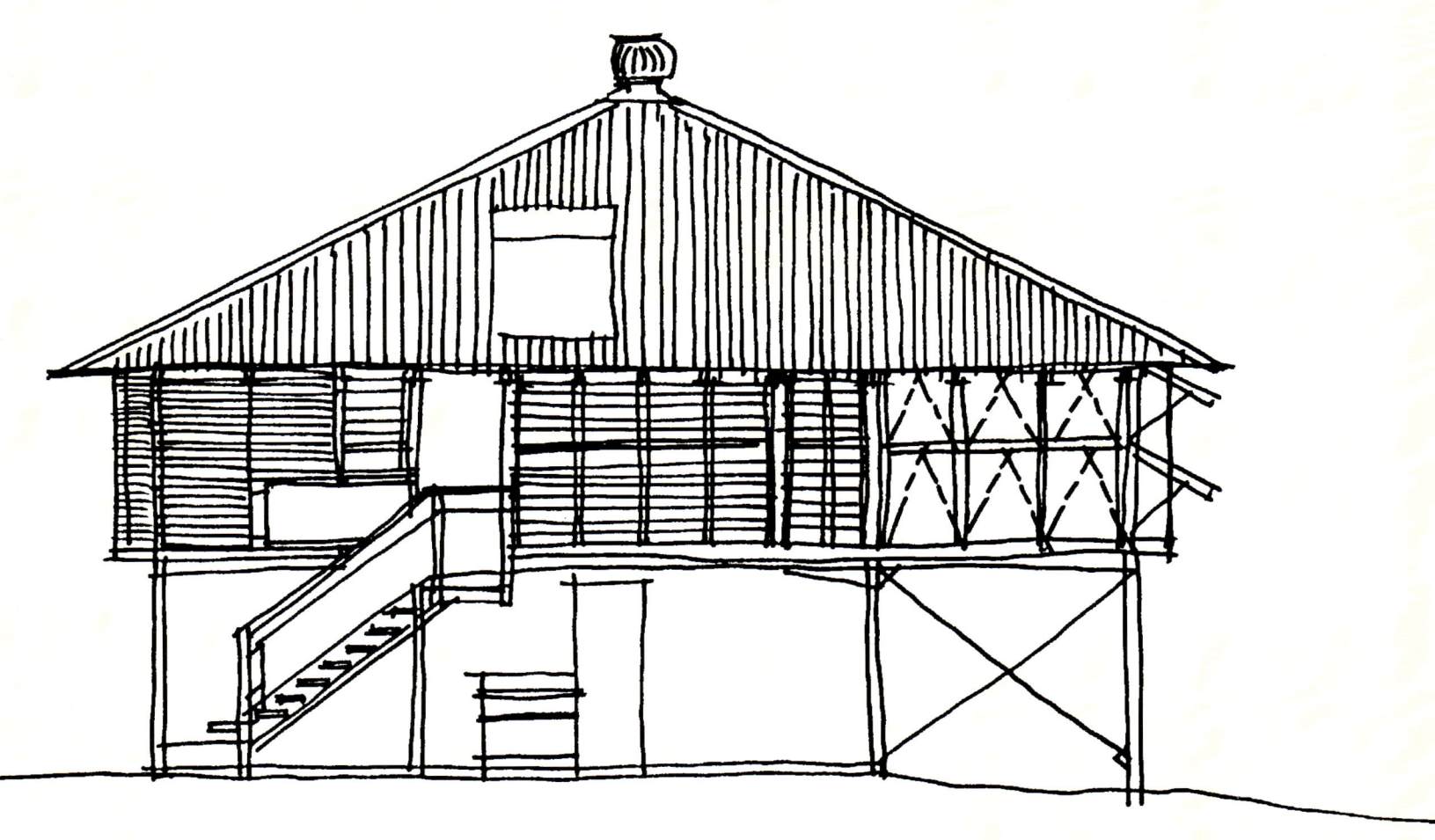

TROPPO TYPE 5 1990 ELEVATION

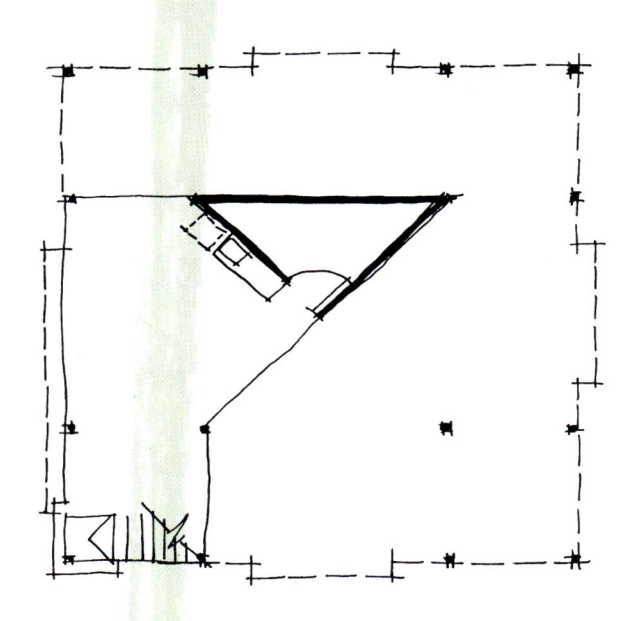

TROPPO TYPE 5 1990 GROUND FLOOR PLAN

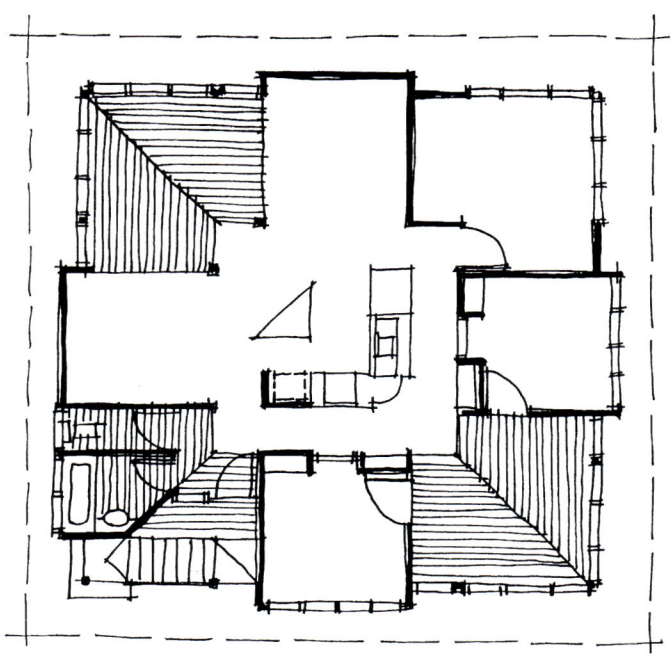

TROPPO TYPE 5 1990 UPPER LEVEL PLAN

TROPICAL HOUSE, PALMERSTON 1990

In 1990, the Department of Lands and Housing ran a Tropical House Design Competition with the intention of producing a suite of three bedroom house designs suitable for public housing in the Top End. The 'Pyramid Roof' design by Troppo and Speargrass Architects won first place, and it was constructed as a demonstration house in Palmerston, one of Darwin's recently established satellite suburb/towns. An elevated steel-framed house with walls of corrugated iron, the Tropical House (as it became known) had a square plan with a pyramid roof that had a spherical spinning vent at its apex. Designed on a composite system of post and beams, and laid out on a 2.7 metre grid from a 5 metre central square, the disposition of rooms delineated a cruciform with the kitchen at the centre with two bedrooms, dining and living alcoves forming the four arms of the cross. Diagonally located, living and sleeping verandahs opened directly off the kitchen. These enabled excellent cross ventilation, and also the sense of being at once outside yet inside. All the spaces were contained beneath the all-encompassing pyramid and within the perimeter of the square plan. The second diagonal was occupied by entry/bathroom and the main bedroom. The effect was like living on a verandah. The whole house was described by Troppo as being essentially made up of 'shuttered verandahs'.

Tropical Housing for the Forces

Success with the Tropical House led to major commissions from the Defence Housing Authority (DHA) in the early 1990s for housing at Palmerston, the Larrakeyah Army Barracks, and the naval base, HMAS Coonawarra. Darwin and the Top End experienced a minor boom in construction as Australia's defence forces were consolidated in the region. Yet again in a cycle of which Troppo became an intrinsic part, it was government sponsored housing rather than the speculative market which was leading the way in responsible tropical house construction. All the gabled and hip-roofed houses were built to withstand cyclones and were specially sited to catch seasonal breezes. They were shaded as much as possible with wide eaves, window shades and verandahs. Together with a series of other architects, they were commissioned to design individual project houses as well as medium density housing. Troppo's philosophy was, on the one hand, to provide the armed services with a spatial understanding of a suburb from southern Australia. On the other hand, despite the DHA insistence on air-conditioning for many of these houses, Troppo sought to provide single and medium-density housing types specially attuned to the tropics and ideally without the need for air-conditioning.

DHA LARRAKEYAH HOUSING 1992

DHA LARRAKEYAH HOUSING 1992

DHA LARRAKEYAH HOUSING 1992

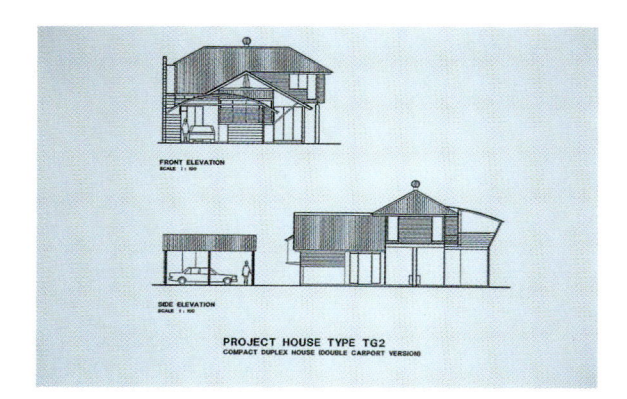

DHA LARRAKEYAH HOUSING 1992 ELEVATIONS

DHA LARRAKEYAH HOUSING 1992 SITE PLAN

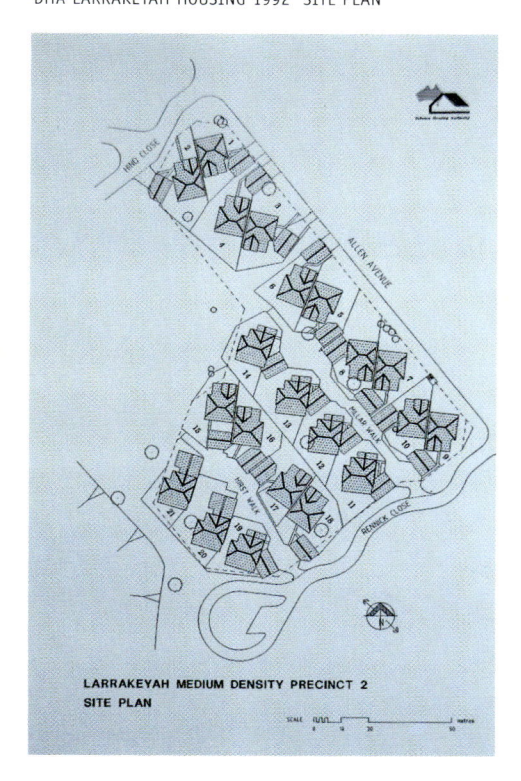

DHA LARRAKEYAH HOUSING 1992

DHA COONAWARRA HOUSING 1992

DHA COONAWARRA HOUSING 1992

MARRKOLIDJBAN OUTSTATION SCHOOL, CENTRAL ARNHEM LAND 1992

Designing for Indigenous Communities

Troppo expanded their understanding of client demands in a completely different way when they began designing for remote aboriginal communities. Learning from architects already practising for some time with aboriginal communities such as Paul Pholeros, and Tangentyere Design (Deborah Fisher) in Alice Springs, Troppo also benefited from Danny Wong's Darwin experiences in aboriginal community consultation. Troppo were expert in paring back technologies, reducing trades and having a small material palette, but in designing for aboriginal people, there were new and constantly evolving lessons. How to design a house for four people that might actually need to house twenty; understanding issues of 'humbugging' where some spaces had to be private and hidden; and how to accommodate the need for their new clients to territorialise or claim the work as their own. There were also more practical matters: the political process of how one was actually commissioned and who by; how to use aboriginal labour on site if appropriate; and how to understand gender roles on site. Troppo found that it was difficult to generalise about a constantly evolving learning process. It was a completely different challenge.

The Marrkolidjban Outstation School (1990-92) had to fit humbly within a series of bush-built timber and recycled corrugated iron structures, one public tap and a small open-sided steel-framed school. It also had to be a place that could fit everyone in the community, not just the children. At the same time, Troppo's school was a natural outcome of their experience with house design in the Top End. Its huge pyramid roof shaded the vast square platform floor and moveable plywood shutters around the perimeter enabled the school to literally breathe. Inside, Troppo specified no paint finish to the plywood wall panels and these were later painted by the local community. The success of this project led to an ongoing stream of commissions from the aboriginal community, including staff housing for Bawinanga Association, Maningrida, Arnhem Land (1991-92) and offices and houses for the Northern Land Council at Borroloola, Ngukurr and Timber Creek (1992-4). As part of the office's expertise, buildings for remote aboriginal communities now account for numerous projects in far flung parts of northern Australia including Gutjangan Outstation School, near Yirrkala, Arnhem Land (1993-94); community housing/aboriginal schools at Kowanyama, far north Queensland (1995-96); as well as buildings in Derby, WA, and Elcho Island, NT.

MARRKOLIDJBAN OUTSTATION SCHOOL, CENTRAL ARNHEM LAND 1992

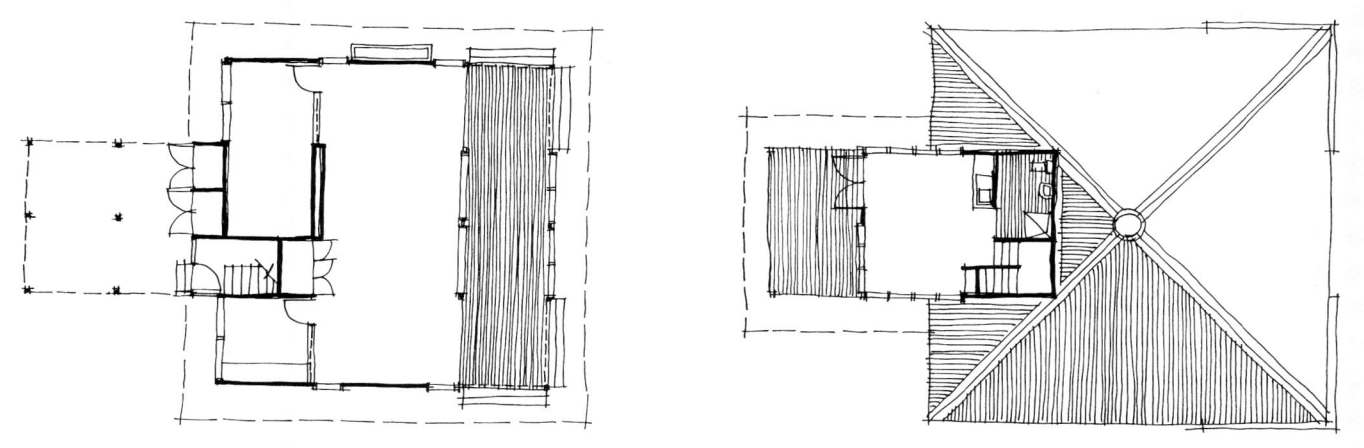

MARRKOLIDJBAN OUTSTATION SCHOOL, CENTRAL ARNHEM LAND 1992 GROUND FLOOR PLAN AND UPPER LEVEL PLAN

MARRKOLIDJBAN OUTSTATION SCHOOL, CENTRAL ARNHEM LAND 1992 (MODEL)

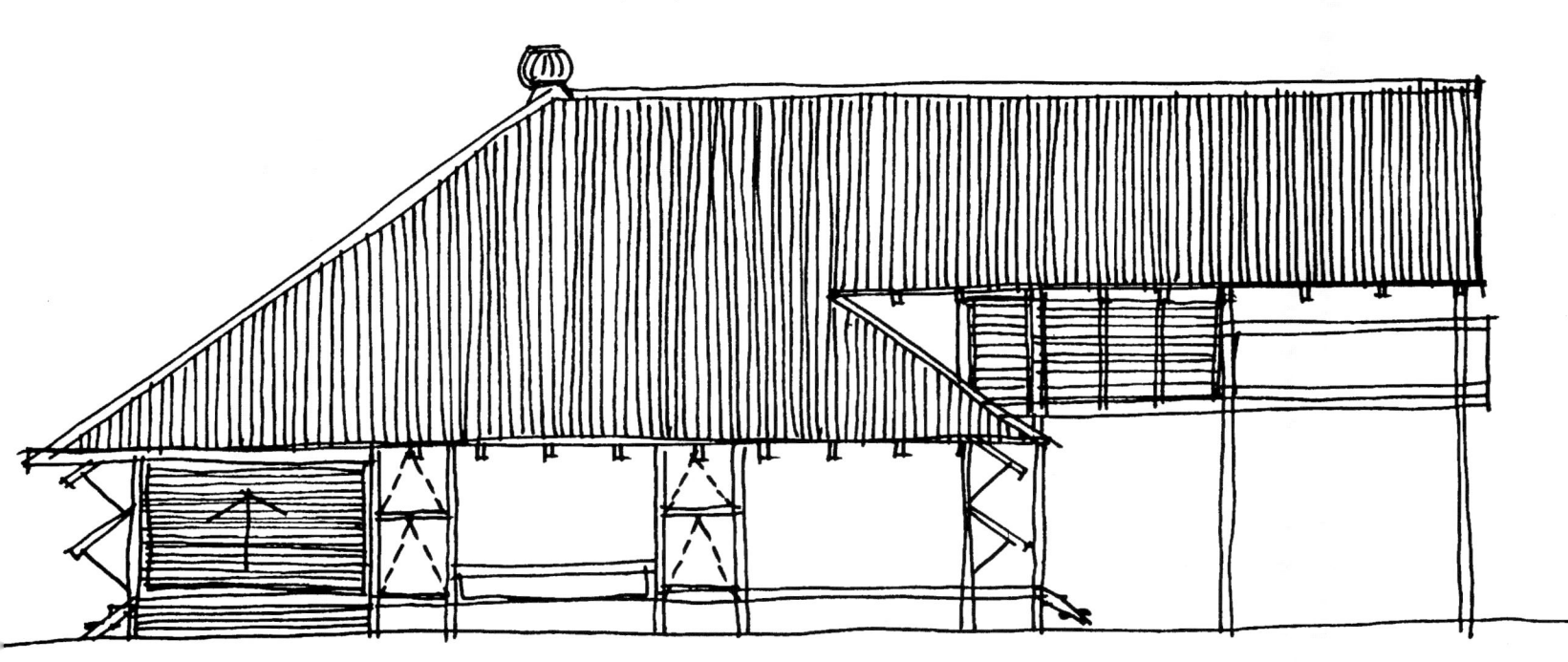

MARRKOLIDJBAN OUTSTATION SCHOOL, CENTRAL ARNHEM LAND 1992 ELEVATION

MARRKOLIDJBAN OUTSTATION SCHOOL, CENTRAL ARNHEM LAND 1992

MARRKOLIDJBAN OUTSTATION SCHOOL, CENTRAL ARNHEM LAND 1992

GUTJANGAN OUTSTATION SCHOOL, EAST ARNHEM LAND 1993

MARRKOLIDJBAN OUTSTATION SCHOOL, CENTRAL ARNHEM LAND 1992

GUTJANGAN OUTSTATION SCHOOL, EAST ARNHEM LAND 1993

INSIDE-OUTSIDE HOUSE, MILIKAPITI 1990

A progression from the Tropical House, the Inside-Outside House has open timber decking within the line of the perimeter wall. It has all the benefits of being outside on the verandah - ventilation, dappled light and space - but with protection from the sun.

INSIDE-OUTSIDE HOUSE, MILIKAPITI 1990

INSIDE-OUTSIDE HOUSE, MILIKAPITI 1990

INSIDE-OUTSIDE HOUSE, MILIKAPITI 1990

DHA RAAF HOUSING, DARWIN 1993

BAKHITA VILLAGE GROUP HOME, COCONUT GROVE 1990

Recognition for Troppo

Troppo continued to design individual houses and inject new formal variations to their work in the early 1990s. The Bakhita Centre Group House, Darwin (1990-94) was an energetic reworking of an existing house damaged by Cyclone Tracy, producing an entirely new house expression. It was a common theme of much of Troppo's other domestic work at the time - the reconditioning of Darwin's old government housing stock that was climatically and spatially inappropriate for modern day needs. Large living verandahs, outdoor kitchens, and spacious shaded entry stairs were added. The meanness of previous housing types was presented with a gift of tropically-inspired space - by the expansion of the house edge to create a layering of shade and entry sequence. Essentially the breadth of the canopy was increased and the shelter density maximised. The opportunity to live beneath an almost limitless roof was offered while forgoing any need for enclosing walls. The upgrades of the RAAF Base housing in Darwin, and DHA housing in Townsville, Qld followed these simple architectural principles.

The multi-level Porcelli/Hazeldine House, Malak (1992-4) was a development of the steel framed treehouse. The design was part of a long term staged process in which the house would, by separate additions, grow into a compound of pavilions. Each building would provide shade, with dense planting creating the sense of a community of buildings within a private jungle. The dense foliage would also act as the first filter to the heat and humidity. By contrast, a beachside house at Nightcliff in Darwin (1993) presented a new theme for Troppo, a bold experiment with a skillion roof, an experiment that would bear fruit in later non-residential projects nearby.

PORCELLI-HAZELDINE HOUSE, MALAK 1992-94

HALL HOUSE ADDITION, COCONUT GROVE 1991

HALL HOUSE ADDITION, COCONUT GROVE 1991

backyard architecture

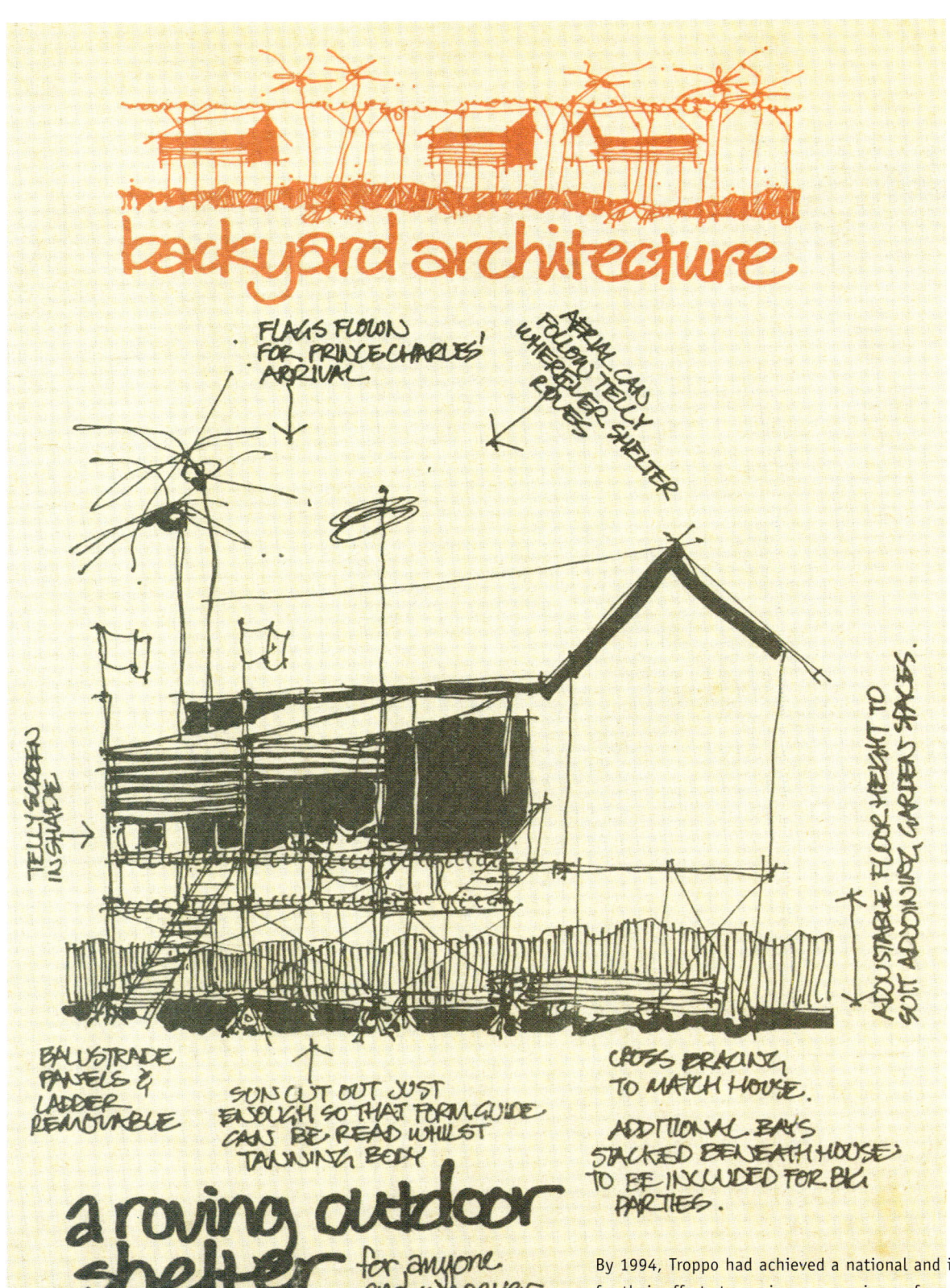

FLAGS FLOWN FOR PRINCE CHARLES' ARRIVAL

ARRIVAL CAN FOLLOW TELLY WHEREVER SHELTER ROVES

TELLY SCREEN IN SHADE

ADJUSTABLE FLOOR HEIGHT TO SUIT ADJOINING GARDEN SPACES.

BALUSTRADE PANELS & LADDER REMOVABLE

SUN CUT OUT JUST ENOUGH SO THAT FORM GUIDE CAN BE READ WHILST TANNING BODY

CROSS BRACING TO MATCH HOUSE.

ADDITIONAL BAYS STACKED BENEATH HOUSE TO BE INCLUDED FOR BIG PARTIES.

a roving outdoor shelter for anyone and whoever.

By 1994, Troppo had achieved a national and international reputation for their efforts to engineer a new image for architecture in the Top End. In 1992, the National Trust of Australia awarded Welke life membership after more than a decade of free architectural advice, and for Troppo's efforts in saving the Burnett houses at Myilly Point. Professional acclaim also came with a flurry of Territory and national awards from the Royal Australian Institute of Architects (RAIA) in 1992. And in 1993, an extraordinary example of national recognition came in the form of a Special Award to Troppo from the RAIA for their diverse and sustained contributions to tropical architecture in northern Australia. It was just reward for twelve years of perseverance in the cause of enlightening a community and a profession.

STOKES ST HOUSE, PARAP 1992

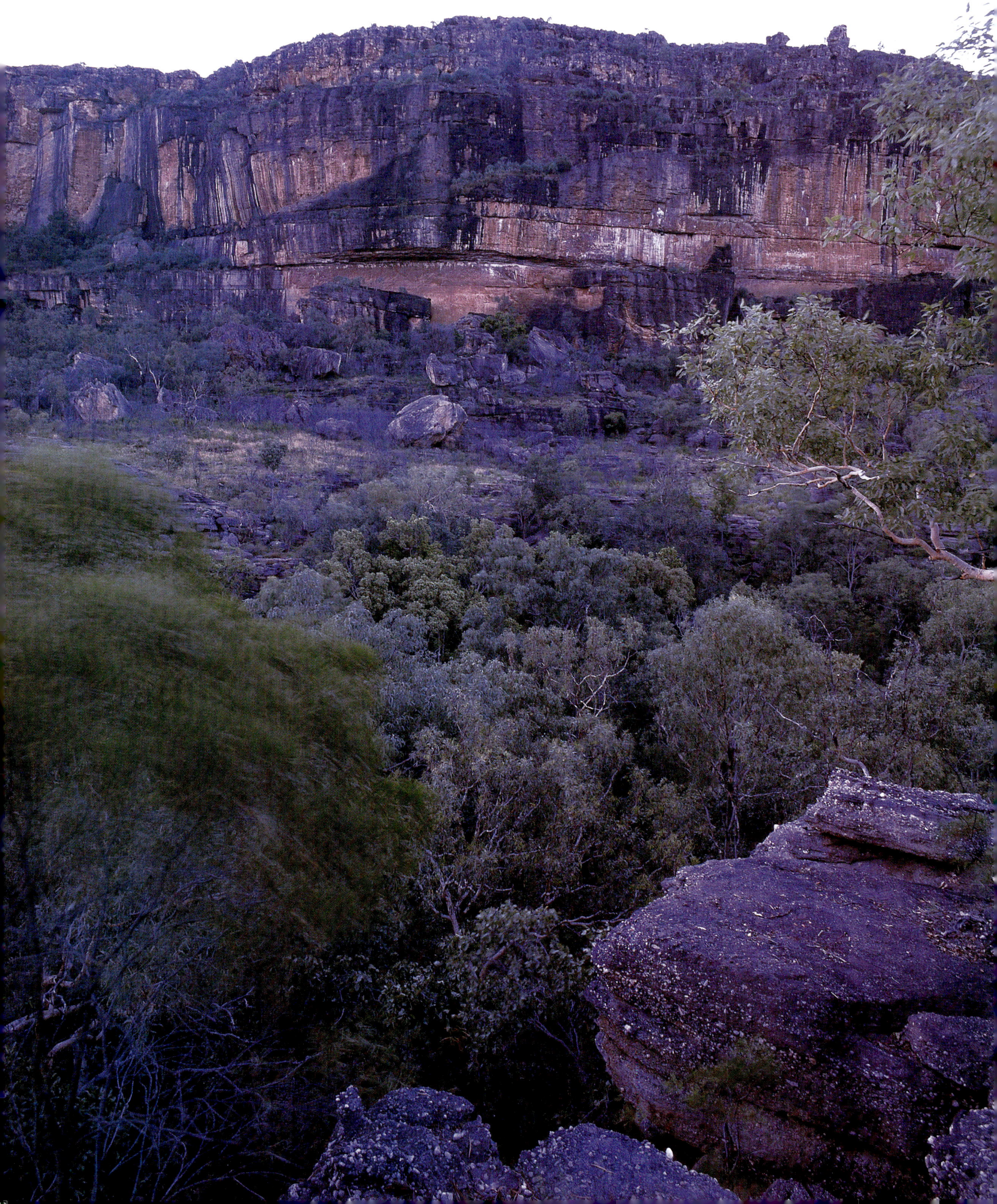

CHAPTER FOUR
NGAD ARRI BOLKNAHNAN

In 1987, Troppo were asked to design buildings at Kakadu National Park, 250 kilometres east of Darwin. It was the beginning of more than ten years of work at one of Australia's most beautiful and culturally significant sites. Kakadu is listed as a place of World Heritage significance, not just for its spectacular landscape, but also because it contains one of the world's richest concentrations of superb rock art. The aboriginal people of the region include the Gagudju (or Ka kudju). In 1912, on an exploratory visit to the remote region, University of Melbourne biologist Professor Baldwin Spencer's name for these people was Kakadu. The name stuck, and it has now been institutionalised in the title of Kakadu National Park. Spreading over 6144 square kilometres of wetlands, plains, dramatic escarpments and home to unique wildlife, and celebrated Aboriginal art, Kakadu is a magical place. The presence of the landscape is enrapturing. The park is owned and managed by the Gagudju people, and to be asked to build there was a special privilege.

KAKADU NATIONAL PARK RANGER HOUSING, MARY RIVER 1989

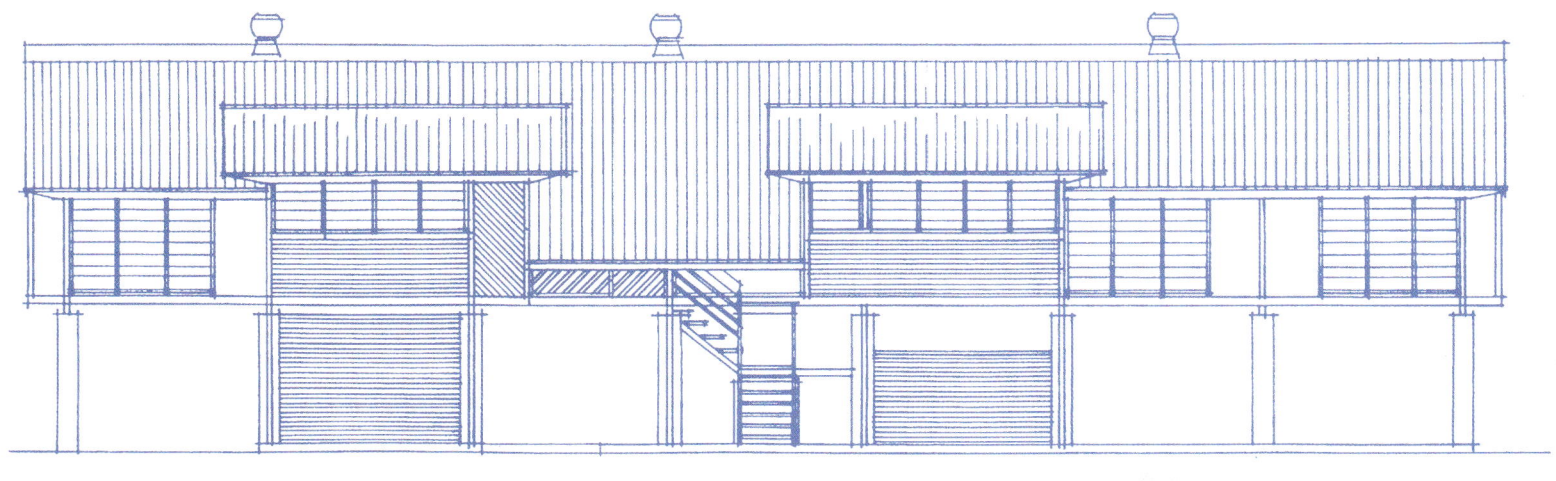

KAKADU NATIONAL PARK RANGER HOUSING, MARY RIVER 1989 ELEVATION

KAKADU RANGER HOUSING

KAKADU NATIONAL PARK RANGER HOUSING, MARY RIVER 1989

KAKADU NATIONAL PARK RANGER HOUSING, MARY RIVER 1989 ELEVATION

KAKADU NATIONAL PARK RANGER HOUSING, MARY RIVER 1989

KAKADU NATIONAL PARK ABLUTIONS BLOCK 1990

KAKADU NATIONAL PARK NORTHERN ENTRY STATION 1990

The first buildings were rangers' housing at Kakadu's Mary River (1989). Elevated (high-set) with stretched linear plans, deep overhangs, louvres and deck-like timber floors inside, these houses were like New Guinea long-houses. They were spindly, lightweight, deep shadowed buildings with open undercrofts. The familiar materials palette of composite steel and timber framing, corrugated steel, glazed louvres and insect screens was Troppo's counter to the dismal massive and climatically inappropriate buildings which had previously been built around the park. At the northern entry station, Troppo provided a leaping cantilever roof as entry symbol, a featherweight contrast to the battered concrete base of the office behind. The Kakadu ablution blocks (1990) were low-set and linear, with softly rounded gable roofs. Slatted timber, silver and watery green coloured corrugated iron, and walls which never touched the ground meant that these very practical slivers of building virtually camouflaged themselves in the bush. The Bird Information Centre (1988) (though less subtle in form and compromised by its need for air-conditioning) employed the same silver grey colours and steely textures. Its siting would be influential for Troppo's later Kakadu commission, the park's main visitor information centre at Jabiru where thousands of tourists would be oriented for their first visit. It was an extraordinary opportunity - the major architectural preface to an experience of the park's own spectacular natural architecture.

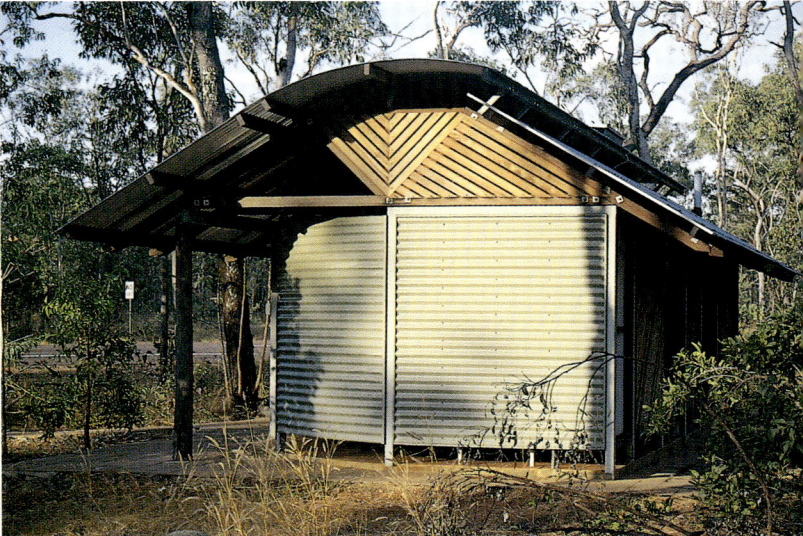

KAKADU NATIONAL PARK ABLUTIONS BLOCK 1990

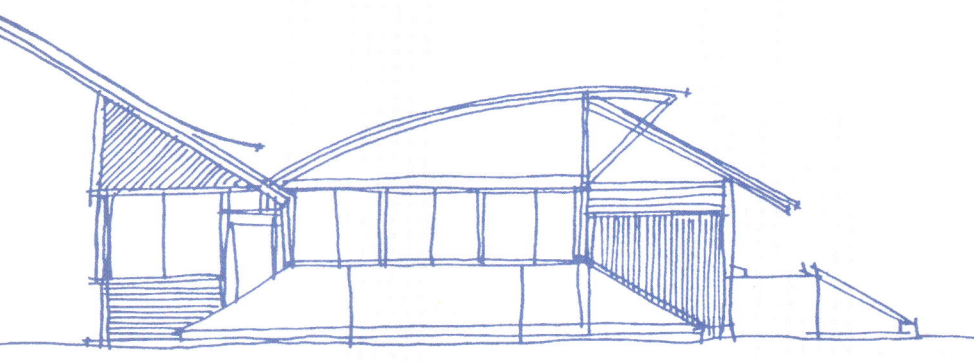

KAKADU NATIONAL PARK ABLUTIONS BLOCK 1990

KAKADU NATIONAL PARK NORTHERN ENTRY STATION 1990 ELEVATION

BOWALI VISITORS INFORMATION CENTRE, KAKADU NATIONAL PARK 1992-94

Bowali

The commission for the Bowali Visitors Information Centre (1992-4) was not automatically given to Troppo. Its importance meant that the Australian Nature Conservation Agency (ANCA, formerly the National Parks and Wildlife Service) and the park's board felt the need to seek expressions of interest. Troppo asked Sydney architect Glenn Murcutt to make a joint submission. They had met some years before in 1983 when Murcutt had seen a National Times article on himself in their shopfront office window. Troppo had long been admirers of Murcutt, and after that chance meeting they invited him to speak to architects in Darwin. They also showed him around many of Darwin's important historic examples of tropically responsive architecture including the 1910 slatted houses which he instantly admired. He measured the spacing of the battens and studied the houses' folding timber shutters. It was an experience that would give sustenance to later projects like his Marika Alderton House at Yirrkala. Murcutt subsequently came back to Darwin on a number of occasions, sometimes as an architect, at other times as a juror, or tourist, but also as a good friend of Troppo.

Troppo and Murcutt won the commission - it was a dream (and also, in some respects, an unlikely) team. One part of the team was a set of practitioners who understood the needs of the makeshift and flexible nature of practice in the Top End and the accompanying demands of its unforgiving climate. The other was an individual architect who had captured one poignant way of understanding the Australian landscape and whose work was highly wrought and intricately planned. The collaboration also included the traditional owners of Kakadu. They formed a crucial part of the design and the building process. The result was a building where the landscape and the message of the park transcended the design philosophies of the individual architects.

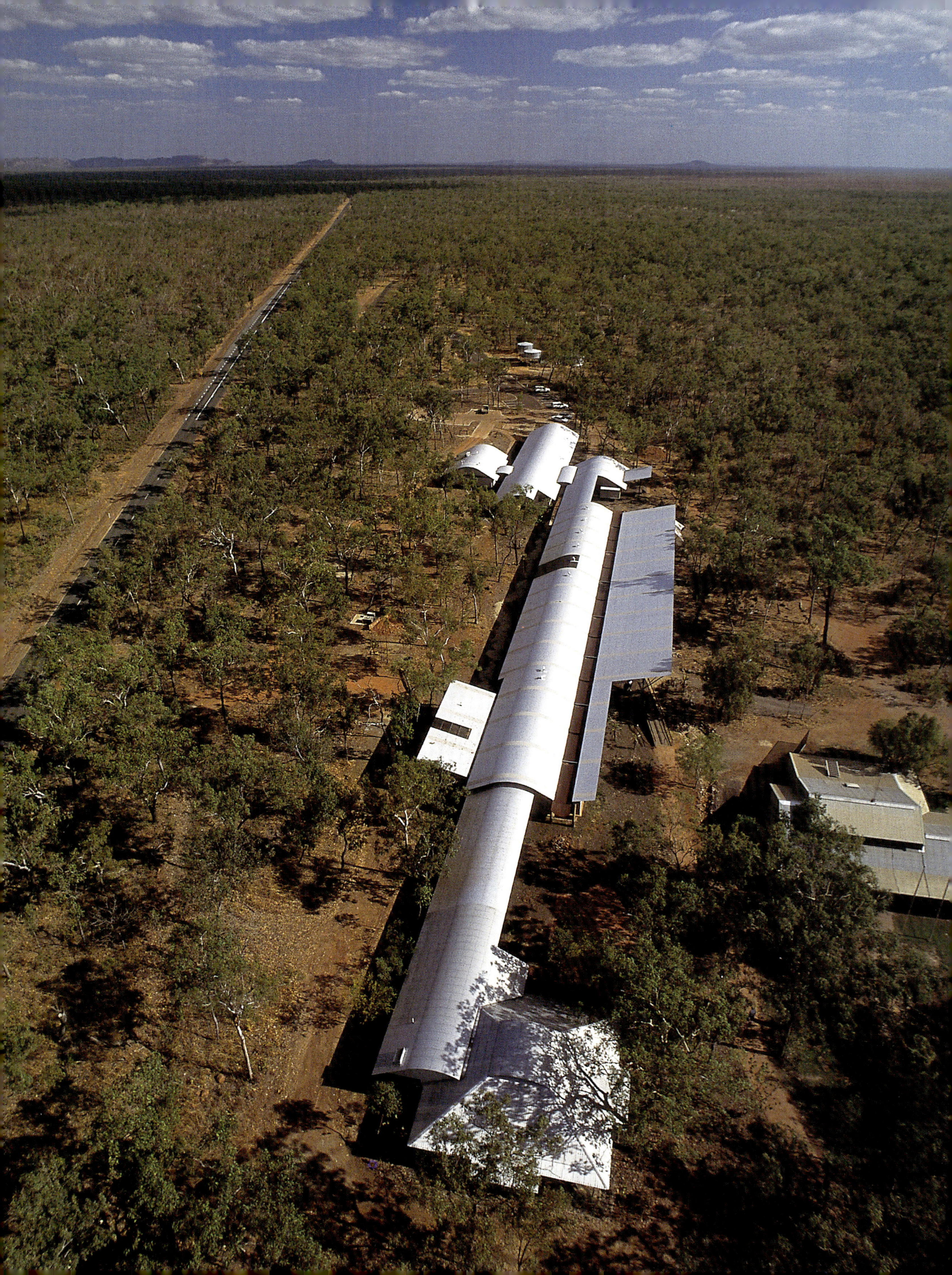

ALL PHOTOS: BOWALI VISITORS INFORMATION CENTRE, KAKADU NATIONAL PARK 1992-94

Confronted with an expanded brief and a group of existing buildings which constrained the master planning of the savannah bush site, it was decided to reorganise, demolish and extend various buildings on site. The Bird Information Centre became the datum point for a linear outdoor corridor deck that would connect the earlier building with a new group of structures - a visitors centre and rangers' offices. However, from the bus park, visitors would be encouraged to wander through a bush track to arrive at the centre. The approach was not on axis - but on the diagonal. With the advice of Big Bill Neidge, a prominent member of the Gagudju, the diagonal or oblique entry was emphasised as the proper approach to natural sacred sites. This was potent advice - a resultant dynamism was gained from these deliberately orchestrated oblique views.

The form and disposition of the visitors centre was like an exploration of the park. It was called Bowali after the Mirarr Gun-djeihmi clan's name for a nearby creek. The building celebrates the collection of water with a massive internal gutter that spills the water onto the rocks below during the monsoon. The oblique approach is a gentle arrival. One comes upon Bowali, 'sidling up to it' as Murcutt suggests. The obliqueness helps to understand the building's functional divisions when arriving at the front deck. There are choices to be made after one moves from the heat of the bush to the cool of the verandah. This is no conventional

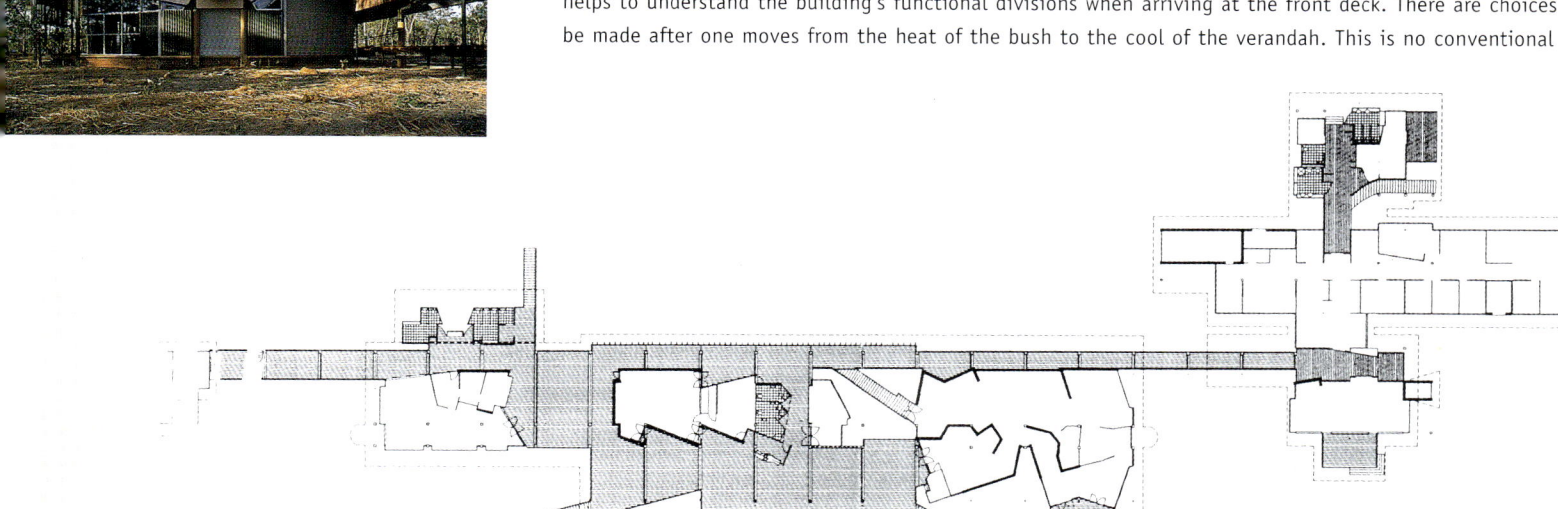

BOWALI, PLAN

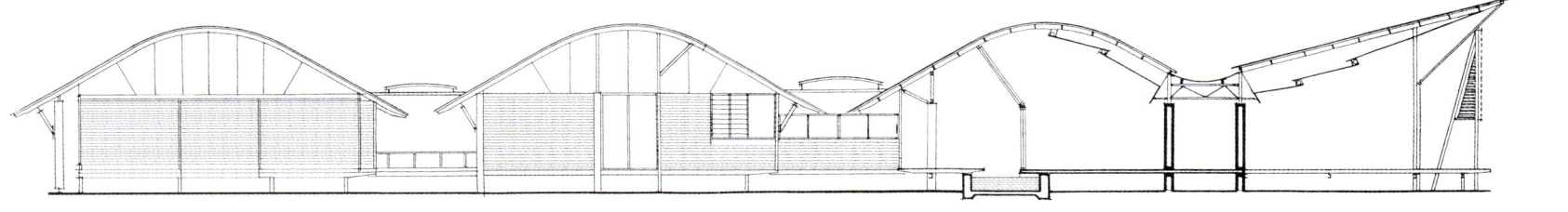

verandah - the roof leaps up to the sky like the undercrofts of Kakadu's caves beneath the escarpments. The underside of the skillion is treated like the layering of rock, a series of dark stained plywood sheets that emphasises the laminated nature of the canopy. One immediately retreats into the cave, towards the mass of the building, a spine of rammed earth in the same rich ochre colour of the rocks of Kakadu. The verandah is also shaded by an undulating angled screen of battens, which provides a constantly changing dappled light - the splintered light of the Australian bush that dissolves a building's form.

Along Bowali's length on the rear and front decks, or internally, one always senses evocations of the landscape. The rust coloured earth walls are the same height as the ant-hills that one sees on entering the park. The plywood panelling could almost be sheet bark. The building is completely open at three points along its length: first, the dark shaded terrace of the café; then the open breezeway halfway along the building; and then finally across an internal billabong crossed by a bridge. Each of these openings will catch any zephyr, any breeze to cool the air, while beneath one's feet there is decking, itself always ventilating. Roll-down blinds and semi-translucent roofing over the breezeways achieved by corrugated fibreglass and perforated metal add to the building's multi-layered and diaphanous quality.

Bowali is a captivating structure. From the approach side, the edge of the skillion roof is so sharp as if to disappear. With the shield of battens below, it is as if the building does not exist, but is simply a collection of sticks. From the other side of the building, the long horizontal of the gable roof of silver corrugated iron provides an heroic view of the timeless rural shed, and in a certain light, it becomes simply a brilliant reflective sky. Unsurprisingly, Bowali earned universal acclaim for the architectural partnership.

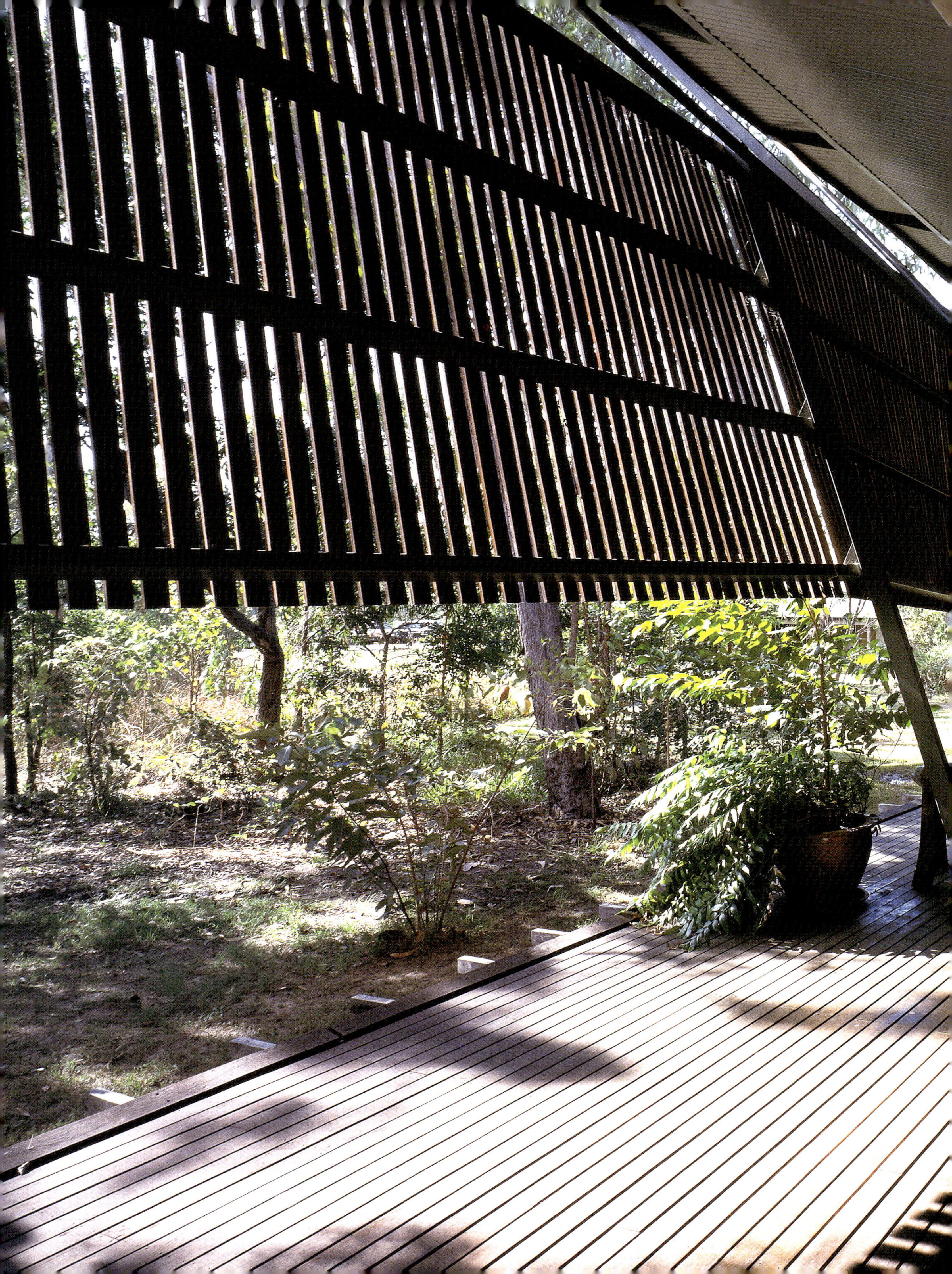

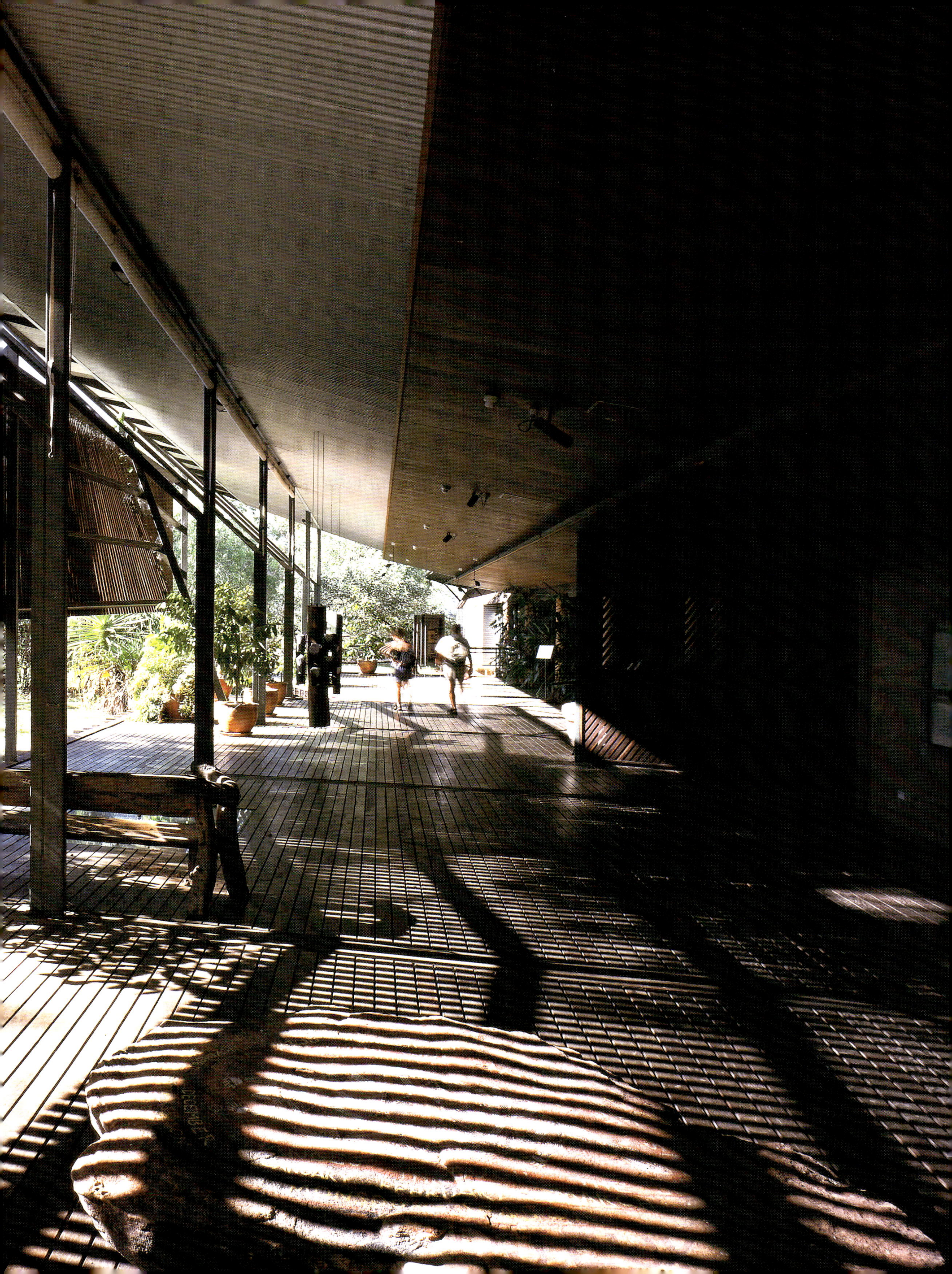
DECEMBER

KAKADU NATIONAL PARK RANGER HOUSING, JABIRU 1995

KAKADU NATIONAL PARK; THE LOVE TENTS, JABIRU 1995

More pragmatic, but no less innovative, commissions followed for Troppo at Kakadu. The need to provide more housing for rangers at the Old Contractors' Camp, Jabiru (1995) for the Djabulukgu Association resulted in the aptly nicknamed 'six-packs' of relocatable steel framed and corrugated zincalume-clad housing units; and 'hard ones' - tiny gable roofed dongas. The 'love tents', one of Troppo's most joyous inventions, are expanded metal-clad and steel framed sleeping boxes elevated above the ground, roofed by stretched paraboloids of white shade cloth, and topped by spherical ventilators. Echoing the timber framed and drooping pyramid roof tents of World War II, the love tents form part of a nomadic encampment. The six-packs, hard ones, and the love tents are planned as a community with tiny streets, around water taps and eating shelters, and all densely planted with quick growing (and temporary) tropical palms and shrubs.

Troppo's work at Kakadu is testament to a maturing of the firm's work, not just in terms of refining the elements of a tropical architecture, but also in their response to a challenging client and an overwhelming place. The sense of an immanent itinerancy about the architecture is positive. There is no cloying cuteness, nor is it tasteless *architecture parlante*. Permanence is not essential here. It is almost an affront to the timelessness of the landscape, but Troppo's tin shelters, their elevated hut-like housing, and the Bowali Visitors Centre will be fondly cared for. As with Kakadu itself, one is always aware of the local maxim for this place: Ngad arri bolknahnan - Look after this country.

KAKADU NATIONAL PARK; THE LOVE TENTS, JABIRU 1995

BOWALI VISITORS INFORMATION CENTRE, KAKADU NATIONAL PARK 1992-94

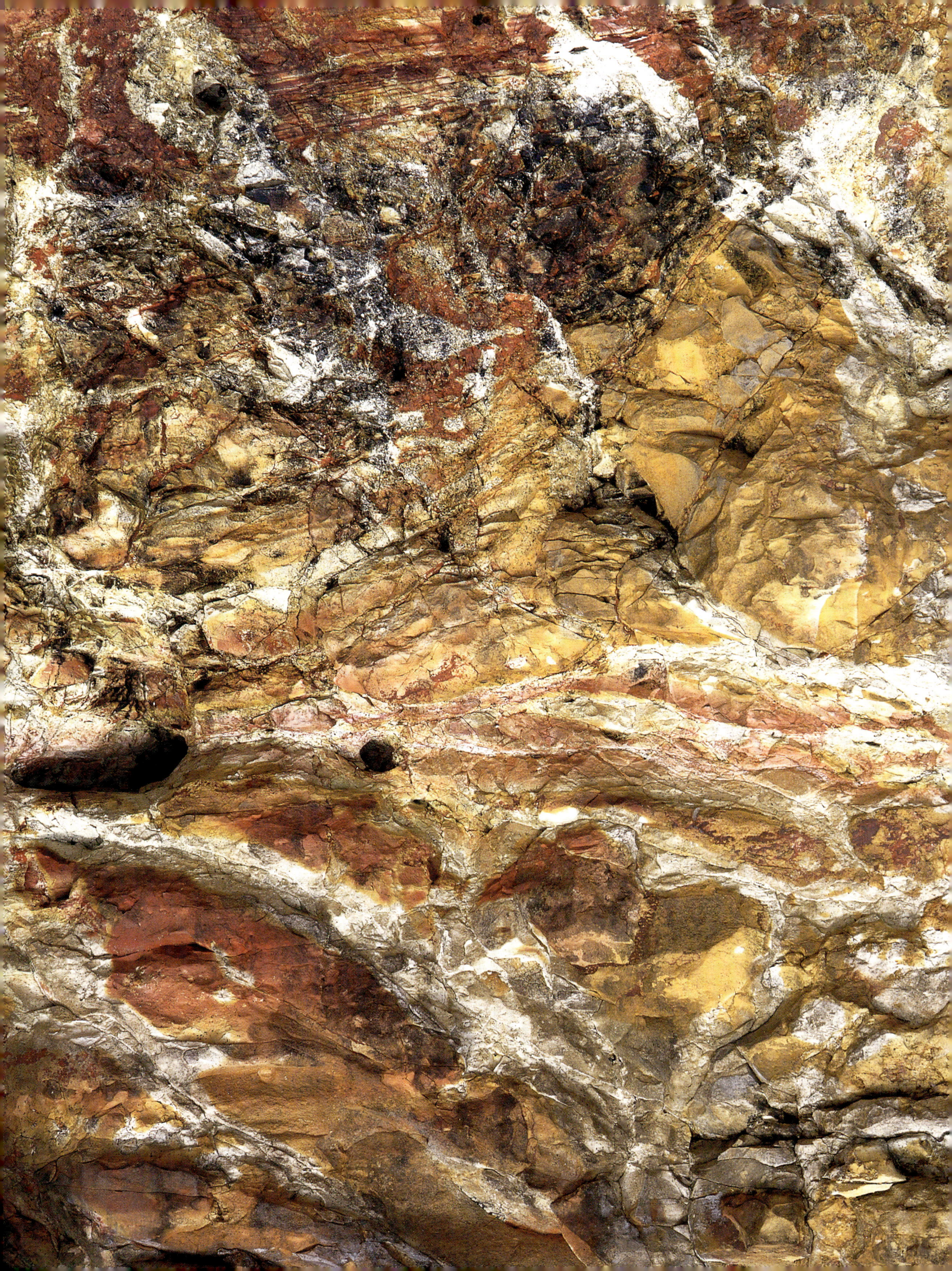

CHAPTER FIVE
PRODUCTIVE AFFLICTION

New and Continuing Themes

Since 1995, Troppo have consolidated and expanded their design repertoire. Ever since settling in Darwin, both Welke and Harris have travelled extensively with their families to nearby regions that had architectural and climatic affinity to the Top End. Places like Bali and other Indonesian islands, Sri Lanka, and Papua New Guinea. Their research on tropical architecture has been continuing and consistent. A trip by the Welkes to Malaysia in 1988 resulted in the Rumah Kampung/Kerr House (1989), a local reinterpretation of the traditional Malaysian house. This infusion of ideas from other places has been complemented by Troppo's continuing interest in documenting Top End architectural heritage. They have undertaken a succession of commissions for heritage surveys and restoration projects including in 1992, a report on the prefabricated Sidney Williams Huts which were built by the Comet Building Company through the Top End during World War II. Troppo had also learnt the benefit of densely planted gardens surrounding their buildings. These gardens greatly assist the shading and cooling actions of their houses. Such influences, ideas and generic principles have resulted in a richer, arguably less free, but more sophisticated design idiom. Troppo's most recent works hint at the idea that they and their architecture might be there to stay in Darwin. They will not, as has been the tradition, just work and pass on through.

RUMAH KAMPUNG/KERR HOUSE, HOWARD SPRINGS 1989

RUMAH KAMPUNG/KERR HOUSE, HOWARD SPRINGS 1989

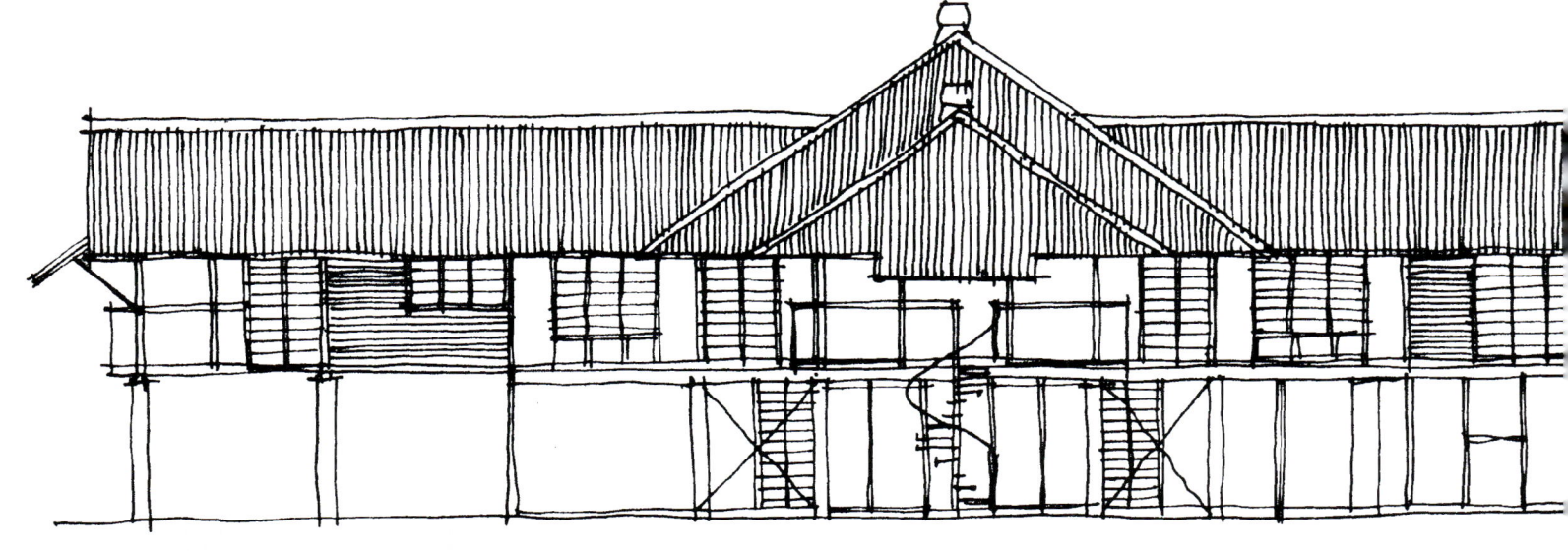

KINGSLEY-KHOO HOUSE, BERRY SPRINGS 1995-97 ELEVATION

Given the late 1990s proliferation of unattractive high and medium rise mass-concrete and block work housing apartments buildings in Darwin, Troppo's Kaiplinger Apartments (1995) are a refreshing alternative. While admittedly low rise and low density, the disposition of the two square planned forms, steep pitched roofs, and spacious verandahs behind an existing surgery means that density comes through an idea of the compound and multi-level tree-house rather than via the slab block or tower. Doing away with lifts and air-conditioning has also meant a more ecologically responsible solution. The idea of the compound also informs Troppo's many recent house additions in suburban Darwin. The Wyville-Smith House (1996) is a 1960s D-series Government house that is long, linear, elevated and louvred. Troppo added an outdoor kitchen and dining deck with a quirky fly-away roof. A swimming pool has been inserted, and enclosed by a bedroom/bathroom pavilion. This completes a spacious compound, a community of airy structures that breathe. The bathroom is open on three sides, reminiscent of Bali. It is not an efficient spatially compact 'machine', but a place to celebrate bathing.

At Berry Springs, in semi-rural savannah, the Kingsley-Khoo House (1995-97) has been designed like a tropical colonial plantation house with spreading linear wings and projecting central deck overlooking the landscape. The deep eaves, the sunshades within the gables, the spindly members and transparency have other allusions, of grander traditional timber community houses of Southeast Asia, of a permanent version of the Macassan huts. These hints at exotic reference are not trite, but always evoked within Troppo's pared back aesthetic. This is a house that will grow organically like its farm, and which will, eventually, be surrounded by beautiful and very large grounds. Back in suburban Darwin, the Copeman House (1996) flirts with more direct reference to Asia. Its pavilion roofs have warped upturned corners, at once referential of Chinese temples but also indicative of Troppo's increasing experiment with the formal possibilities of corrugated iron. At the Welke/Bonney House (1996), the addition of a similar roof with warped corners creates a temple-like draped canopy over the main bedroom. Open on all four sides, this is a bedroom in the treetops above a house which has been added to progressively, and filled with Fran Welke's collection of furniture from Southeast Asia and objects from their travels. It has been growing like the incremental additions of Geoffrey Bawa's houses in Sri Lanka. It is yet more evidence that a Troppo house is an open framework for inhabiting. There is no precious line of puritanism about appearance. It is about living in the heat - not surviving in it, but luxuriating in it.

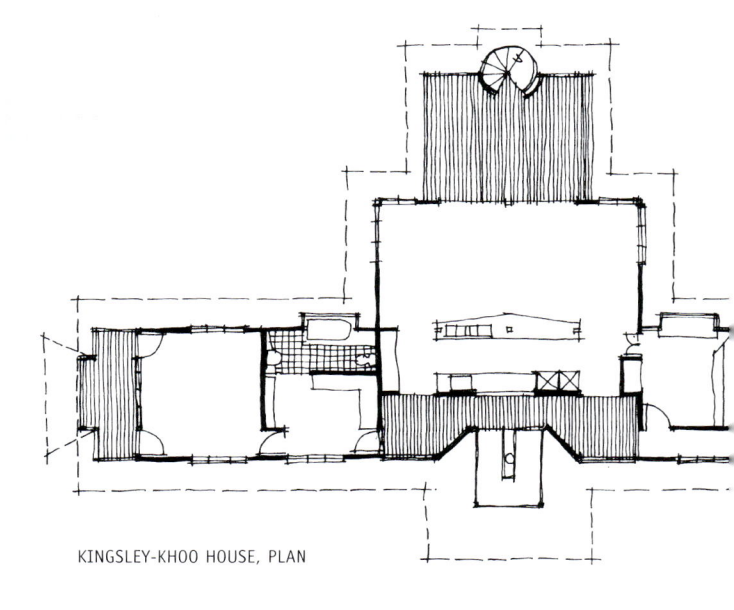

KINGSLEY-KHOO HOUSE, PLAN

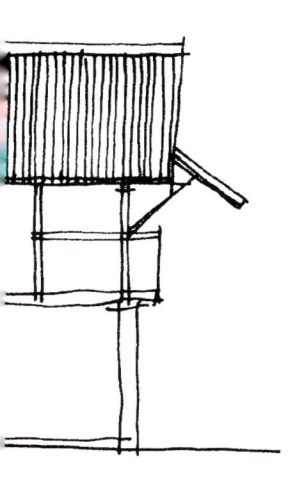

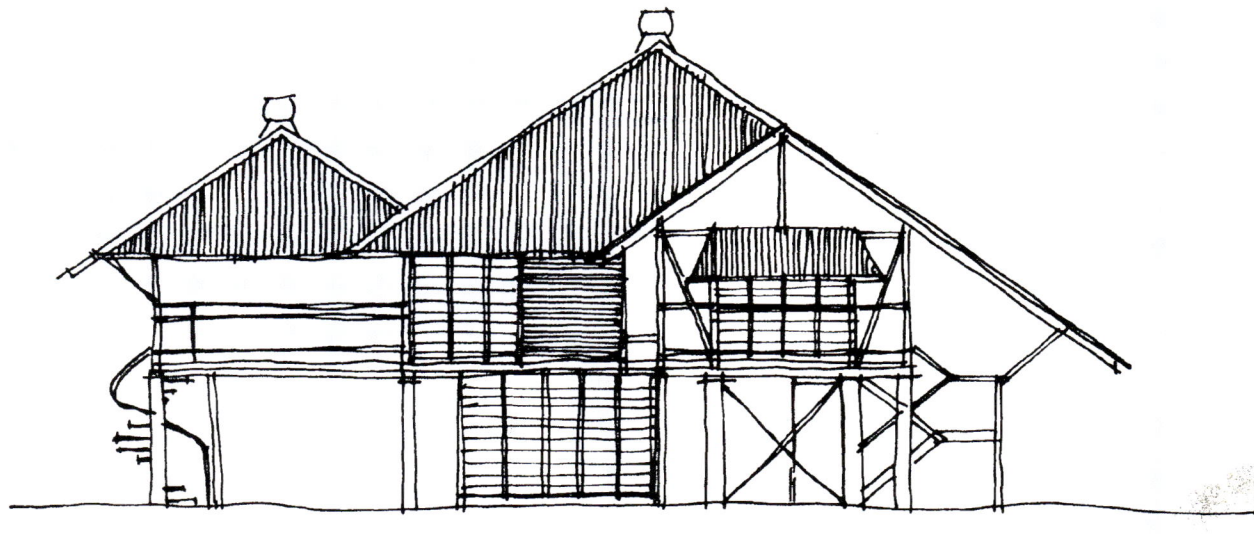

KINGSLEY-KHOO HOUSE, ELEVATION

TOP END HOTEL, DARWIN 1998-99 RESTAURANT

Tropical Commercial

Over its two decades of practice, Troppo has also been designing and building numerous modestly scaled commercial projects, and shop and restaurant fitouts such as Christo's-on-the-Wharf (1992). Their most recent large scale commercial commission, the Top End Hotel (1999) in downtown Darwin was a complete refurbishment (with additions) of a 1960s hotel and bottle shop of the same name. Complemented by pools, landscaped gardens, decks and rocks that ooze water, and with David Lancashire's decorative signage and graphics, Troppo created a series of elegant and upbeat open dining and drinking terraces. There are spaces for beer drinking and stockwhip display, and a massive verandah for fine dining. Gentle skillions, stained Malaysian timbers, Australian ironwood and slabs of porcellenite are mixed with hard-wearing industrial materials. This is Troppo's most adventurous experiment with materials and textures.

TOP END HOTEL, DARWIN 1998-99 BEER GARDEN

TOP END HOTEL, DARWIN 1998-99 MAIN BAR

TOP END HOTEL, DARWIN 1998-99

PEE WEES AT THE POINT, DARWIN 1998

Completed one year earlier, Pee Wees at The Point, Darwin (1995-98) is completely different from The Top End Hotel in setting and intention. Situated amongst lush tropical trees and palms, and looking back across Fannie Bay to the city, this restaurant occupies the site of the former Pee Wee's Camp. This was a loose grouping of corrugated iron clad Sidney Williams Huts originally built to house government construction workers, and which later became the main messing facility for troops located at East Point in the build-up to World War II. (Darwin was an important airbase for the Allied Forces in WWII and was the only Australian city to suffer sustained bombing damage). Recognising the heritage significance of the existing prefabricated steel huts and their original disposition across the site, Troppo retained some of the huts but built an entirely new structure on the foundations of one that had been demolished. Instead of mimicking the huts' gabled forms, Troppo designed the café restaurant/function room as a transparent skillion-roofed pavilion. It is separated by a breezeway from the kitchen and amenities block whose archetypal shed form evokes the history of the buildings nearby. Glass louvres, translucent polycarbonate and a series of glazed steel doors which fold away to entirely dissolve the south walls, combine with reflective corrugated zincalume on walls, ceilings and soffits to produce an airy succession of terraced spaces. At Pee Wees, Troppo perfected their vision for the non-air-conditioned public space in the tropics. It is a lighter more delicate version of the open lounge/bar of Stephenson and Turner's Darwin Hotel (1939). It also celebrates the thinness of structure and the strength of simple gestures to frame views, enclose space and highlight the threshold underfoot. Slabs of hoop pine plywood flooring and timber decking provide warmth of colour and sound before one moves to the grass lawn, and the rocky shore of the bay.

PEE WEES AT THE POINT, DARWIN 1998

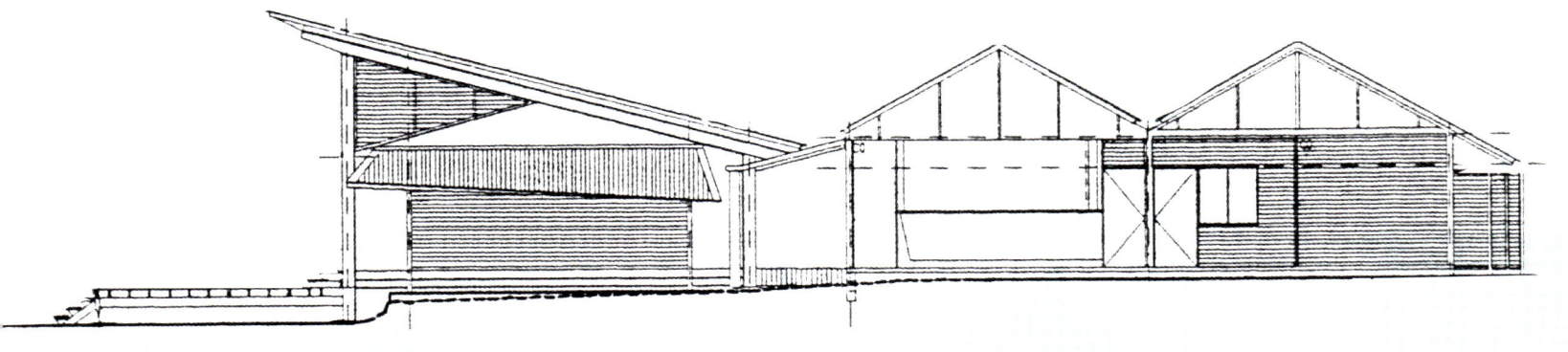

PEE WEES AT THE POINT, DARWIN 1998 ELEVATION

PEE WEES AT THE POINT, DARWIN 1998

ALL PHOTOS: PEE WEES AT THE POINT, DARWIN 1998

75

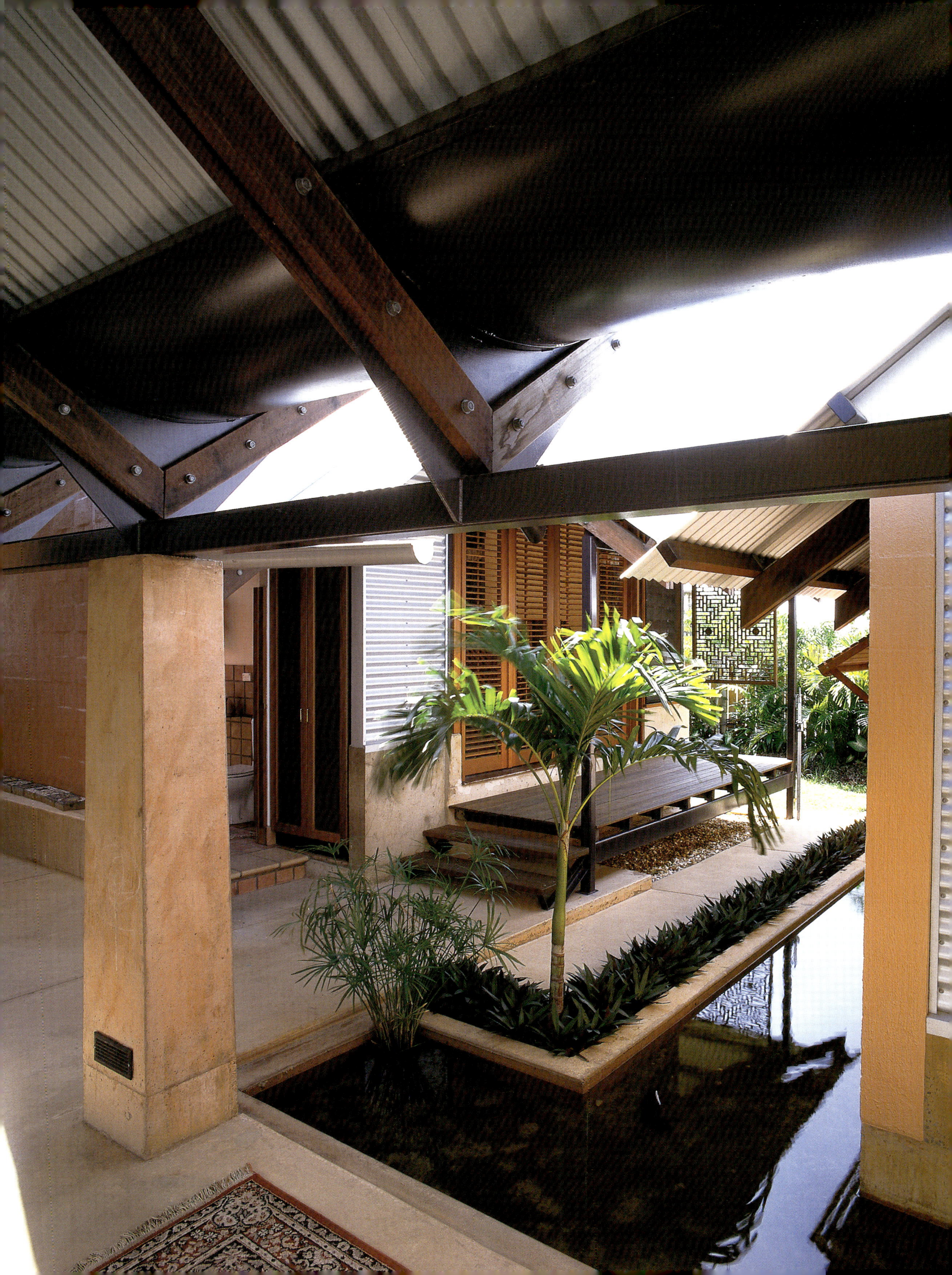

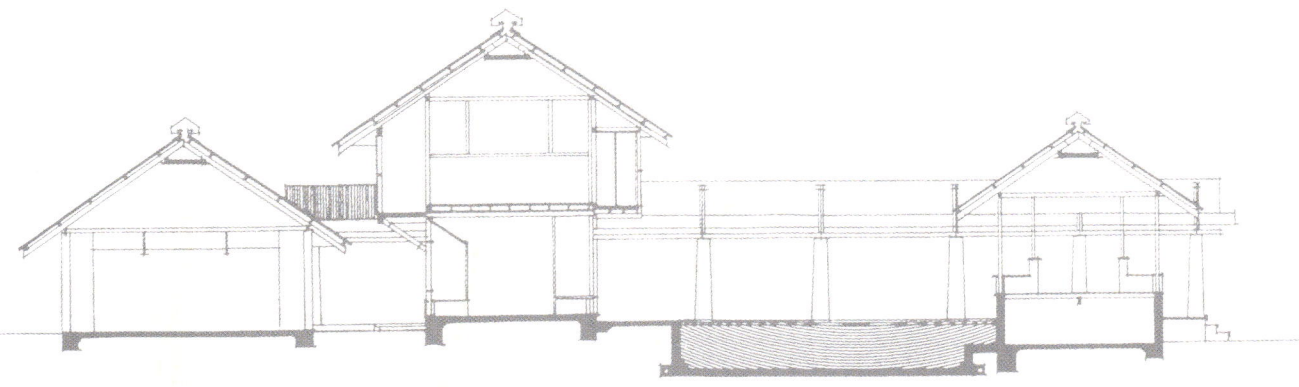

THIEL HOUSE, CULLEN BAY 1998 SECTION

The Thiel House (1998) built right on the water at Darwin's Cullen Bay is Troppo's most recent attempt to understand the idea of the house in the Australian tropics, this time as a traditional Balinese courtyard-house compound. In a progression from previous work, the Thiel House makes use of mass, in the form of tapered concrete piers and terraced floor slabs; and water, through the use of pools and ponds, the collecting of the rain through a giant internal gutter (an 'elevated river' in the wet), and a constant relationship to the views of the adjacent bay beyond. Designed across the entire area of its site as a series of five open pavilions linked by a circulation spine and enclosed garden spaces, this house - like those of Geoffrey Bawa in Sri Lanka and Kerry Hill in Bali – is a masterful amalgamation of Eastern and Western ideas. Their clients, enamoured of Southeast Asian art and architecture wanted Troppo to evoke an atmosphere of oriental languor. The density of the new Cullen Bay subdivision meant that Troppo had to use solid walls and timber screens, a different material language to their normal palette. In doing so, Troppo ventured into an entirely new area of experiment. While the practice is celebrating twenty years of existence, it is clear that no closure to its activities has been announced. Instead, Troppo appear insatiable – they have just entered another phase of 'going off', of 'going troppo', and finding that it continues to be a productive affliction.

ALL PHOTOS: THIEL HOUSE, CULLEN BAY 1998

ALL PHOTOS: THIEL HOUSE, CULLEN BAY 1998

THIEL HOUSE, UPPER FLOOR PLAN

THIEL HOUSE, ELEVATION

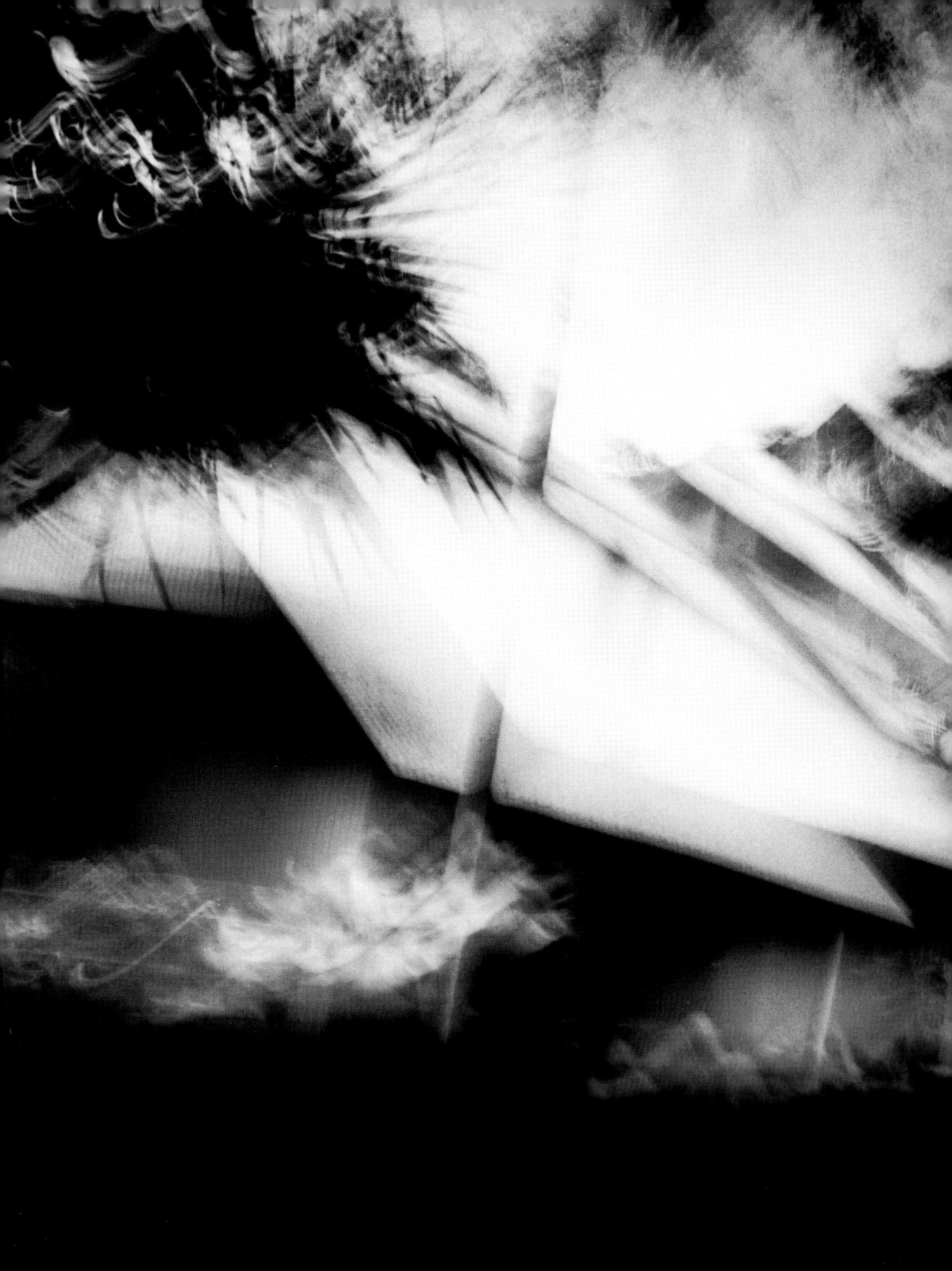

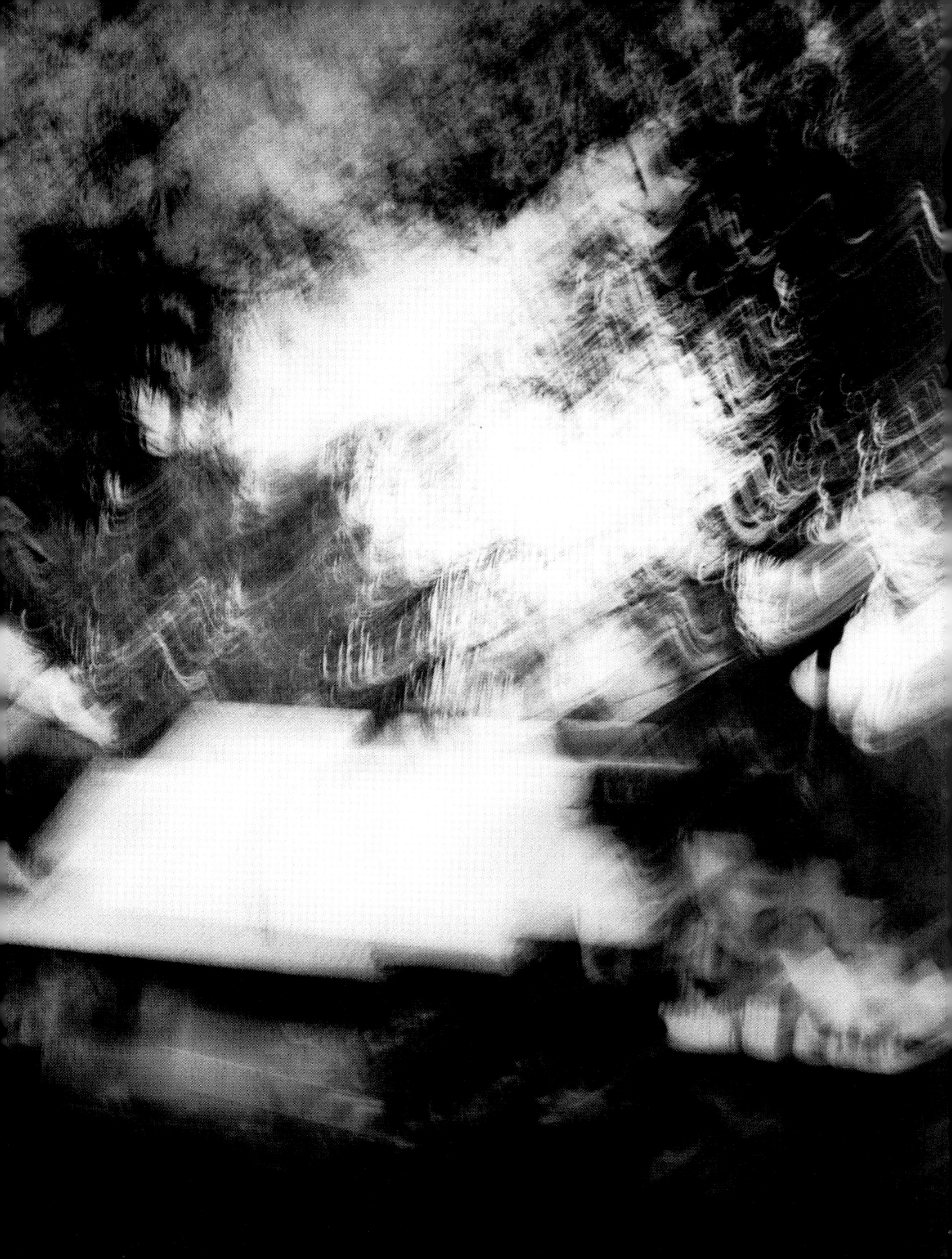

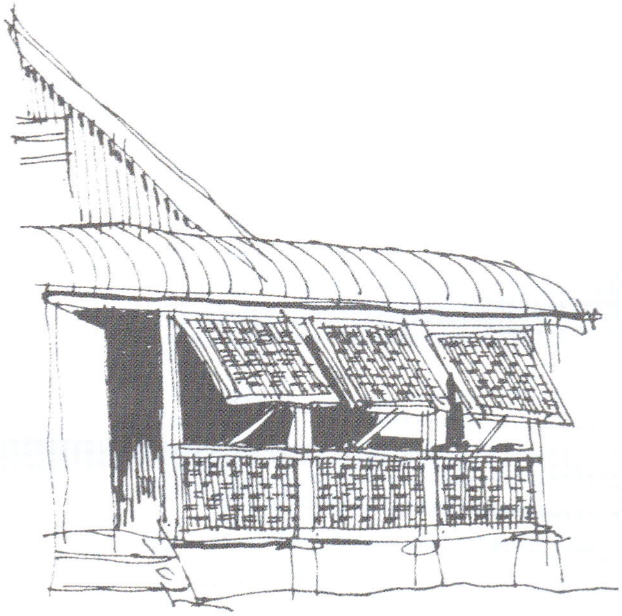

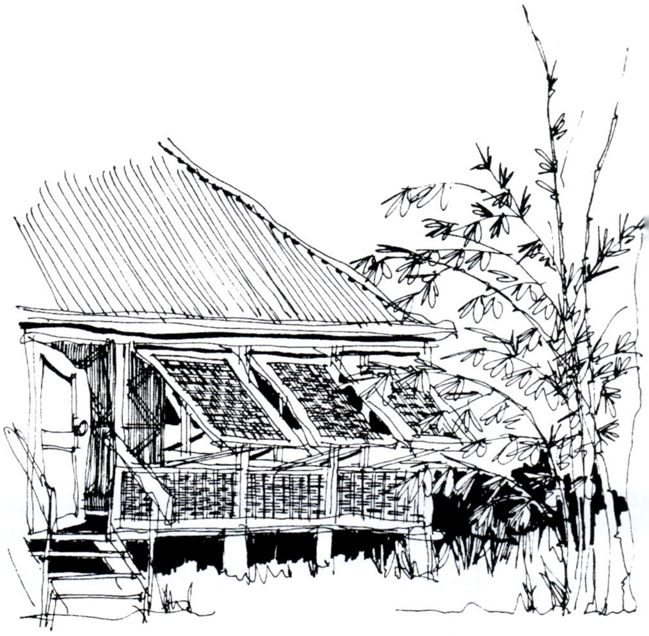

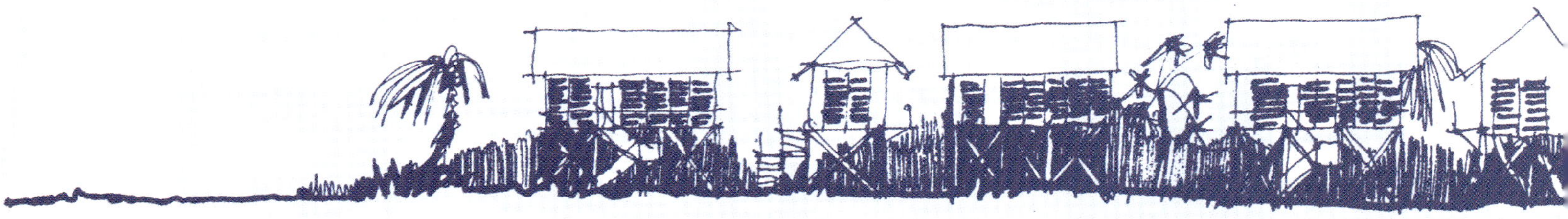

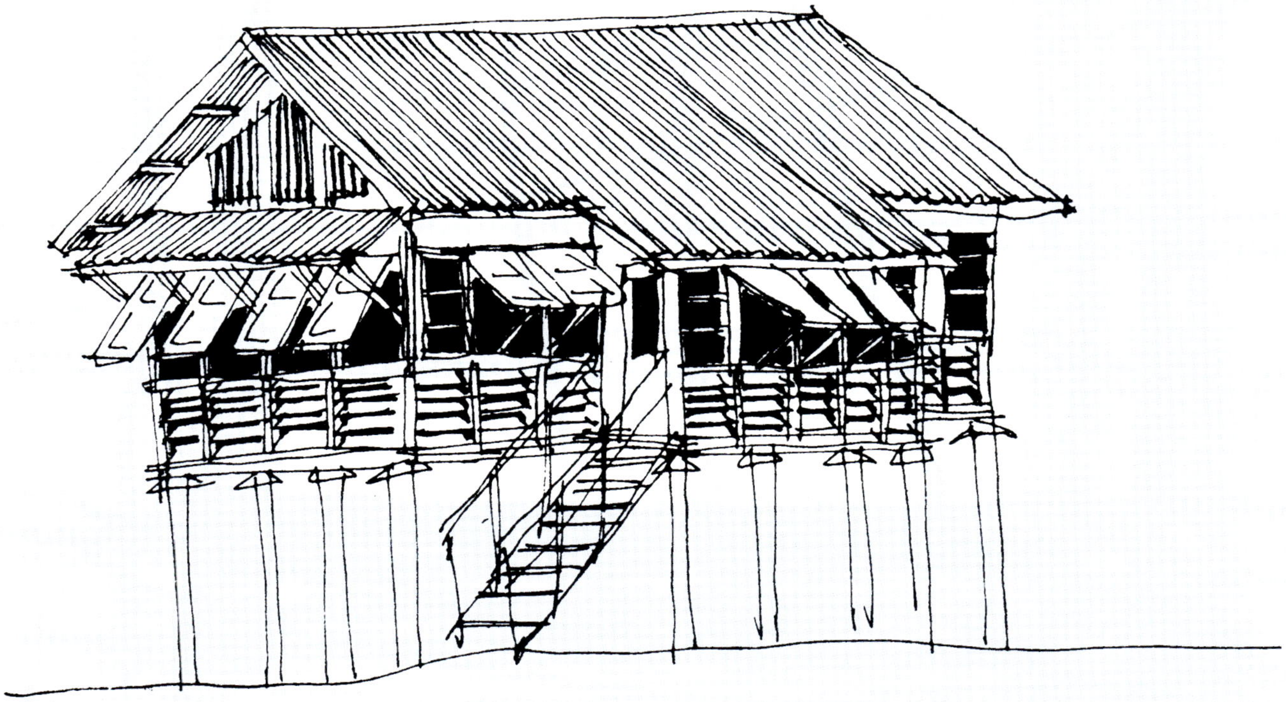

CHAPTER SIX
TOWARDS A TROPPO ARCHITECTURE

Troppo have developed their architectural responses into a series of thematic constants, a series of ideas which can be passed onto successive employees and to students in the lectures that they give at various schools of architecture around Australia. These ideas do not become a restrictive design manifesto, but form a constantly evolving set of general guidelines – part pragmatic and expedient technique; part phenomenological models, and deeply felt responses to the environment and social behaviour. Ten themes now characterise their work after twenty years of practice.

'Hearing the Rain'

Troppo's architecture acknowledges the elements. The thundering of rain in the 'Wet', and the constant dripping of water off broad eaves is part of the place that is the Top End. There is no point in escaping from or attempting to block out such natural phenomena. Taking advantage of the elements means, for example, acknowledging the brightness and intensity of the sun in the 'Dry' with its obverse – deep shade. To celebrate one element will balance the other.

Transported Materials

Troppo's architecture acknowledges the necessity of imported materials in a place that cannot provide enough natural building materials. The use of corrugated iron, steel framing, plywood timber panels, fibre cement sheet, louvres and the corresponding idea of the building as a prefabricated kit of parts, freely assembled and dissembled, informs their architecture. In the Territory, most workers have welding kits, they can work out of the back of a ute or in their backyard. It means that the welded joint is favoured over and above the bolted connection. With the welded joint you can be efficiently imprecise. Bolted connections, by contrast, imply fine tolerances and exactitude, not always the most economically sensible option.

House as Compound

Troppo plan houses that grow. Often guest houses are added as separate entities. Kitchen decks can be added, bathrooms appear as separate rooms. The house becomes a community of rooms, a community of varying degrees of privacy and openness. Spaces between the structures of the compound become outdoor rooms. This is a hybrid idea of extendable space, the free flow of air, and small spaces made large. It is the idea of the house as village, the house as compound.

Bali Bathroom

In many of Troppo's houses, the bathroom becomes a celebration of the act of bathing. While washing, one faces the landscape and communes with nature. The Bali bathroom is like an open pavilion, a temple of washing where total openness has also practical applications. It eases problems of condensation and mould in a climate where, at certain times, things never seem to dry.

'Nature, in the Territory, looms larger than man'

Landscape and climate are unavoidable in the Top End. Troppo's architecture responds to the need to keep out the rain and the sun, to reduce heat, and to accommodate humidity. It acknowledges the existence of insects and bats. To retreat is pointless. The house should not be seen as an impenetrable barrier and fortress, but simply something to live under. Dense planting becomes a modulator of that omnipresent nature. One can use nature to filter nature rather than combat it with man-made materials and techniques.

The adjustable skin

Troppo's study of the building heritage of the Top End opened their eyes to the benefits of the adjustable skin. The examples of Burnett's louvres, the battens and folding shutters of the slatted houses, roll-down blinds, the inventive use of shade cloth, and the laminating of roof eaves like the canopy of a tree all offer simple lessons. The house becomes an organism of adjustment, its skin like that of humans - an infinitely receptive tissue.

The natural chimney

Troppo's houses eschew the conceit of the flat roof. The gable, or the pyramid roof and its open underside forms a natural chimney. With roof vents, the house becomes its own ventilator. How to maximise that natural effect becomes an architectural preoccupation. Troppo's inventive roof forms are not just an aesthetic exercise, but the means to an ecologically sustainable architecture. Each structure (inevitably one of Troppo's tropical house typologies) is subjected to an overlay of building science and then pared back to an aesthetic, economic and buildable minimum.

The inside-outside house

Catching the breeze becomes a Troppo obsession with the delineation of a perimeter wall. Stepping the floors within to create a space for any zephyr of wind to create cross-ventilation is part of this obsession. Bringing the battened verandah floor inside the perimeter wall to provide an openly ventilating floor creates an inside-outside house. It is a tactic of blurring orthodox distinctions of enclosure.

A house is....

For Troppo, a house is many things. In the Top End, they find that a house is closer to the idea of a cave or an aboriginal shelter. Each time one builds, one builds back to those elemental typologies. In the Top End, one virtually camps, builds a house and 'lives under it'. There is a landing and front steps, small things that increase the house volume to gain intermediary spaces between indoors and outdoors.

The tenth line

Behind all of these strategies is Troppo's notion of the tenth line. When one draws a solid cube in axonometric or isometric, nine lines are required to represent that cube in three dimensions. To draw a tenth line across any of the cube's corners is to immediately imply transparency to the volumetric system. This is an intrinsic design philosophy for Troppo. Their architecture obviates the need for examining the idea of the open frame and potentially unenclosed volumes. Space is extendible and also infinitely adjustable – if one allows the addition of the tenth line.

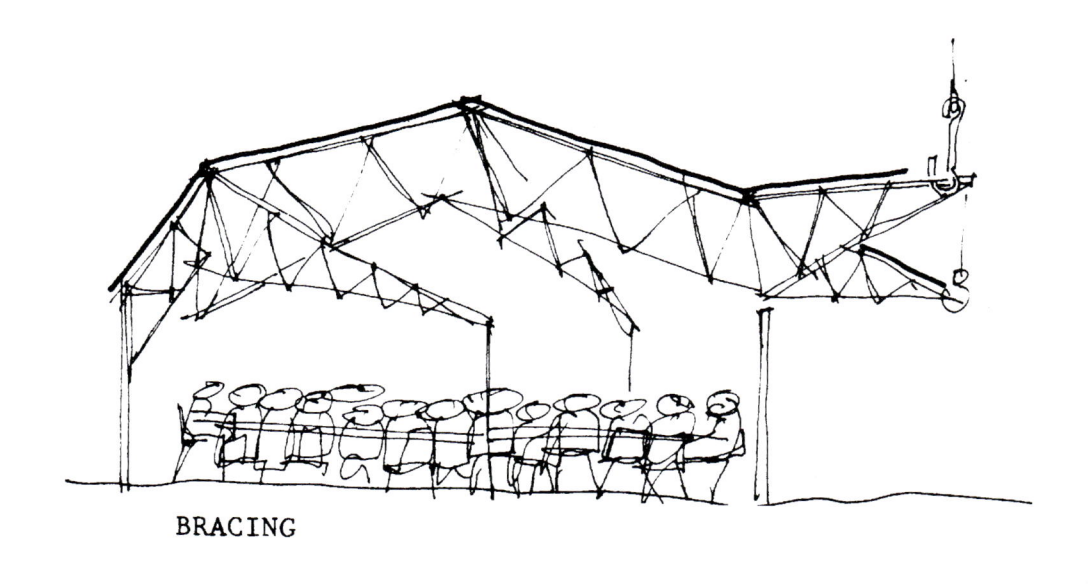

BRACING

RESULTANT UPLIFT.

CLEAT TRUSSES TO POSTS

INCIDENT WIND.

CLEAT POST TO FOOTING

THIEL HOUSE 1998 BATHROOM

IMPENDING PERMANENCE –
TWENTY YEARS IS A LONG TIME

Over the past twenty years, architecture in northern Australia has undergone something of a renaissance. The list of architects from that part of the continent who receive worldwide attention is growing. Troppo can be grouped with Queenslanders Rex Addison, Russell Hall, John Mainwaring, Brit Andresen and Peter O'Gorman, Lindsay and Kerry Clare, and Gabriel Poole. Common to all is the leading role which climate plays in determining the forms of their architecture. It is a quite different preoccupation to architects in southern Australia whose tense manipulations of form and symbolism occupy an altogether foreign architectural position to their northern counterparts. That can only be good for architecture in Australia. When outsiders view Australian architecture, preconceptions abound. Invariably, the image of a romantic primitivism, of a dreamy tent or shed-like house set within an idyllic Antipodean Arcadia is proposed. Troppo's architecture can be seen in this light – for a fleeting moment. There is something tougher in their work, also something a little looser, a bit more flexible, and a bit more tenuous about time. It has to do with their location, the heat, the people and the type of work they do. This is not an excuse but an advantage. It is as if their architecture at times relates to the archipelago lying north of the Arafura Sea and at others to the outback tradition of the rural tin shed, and to the fluid growth of indigenous shelter, a non-constant architecture. It is an open system, indicative of the history of nomadism of the place, but it could also be seen as suggestive of an impending permanence. Troppo, to their own surprise, have become established. They are an institution. After all, twenty years is a long time.

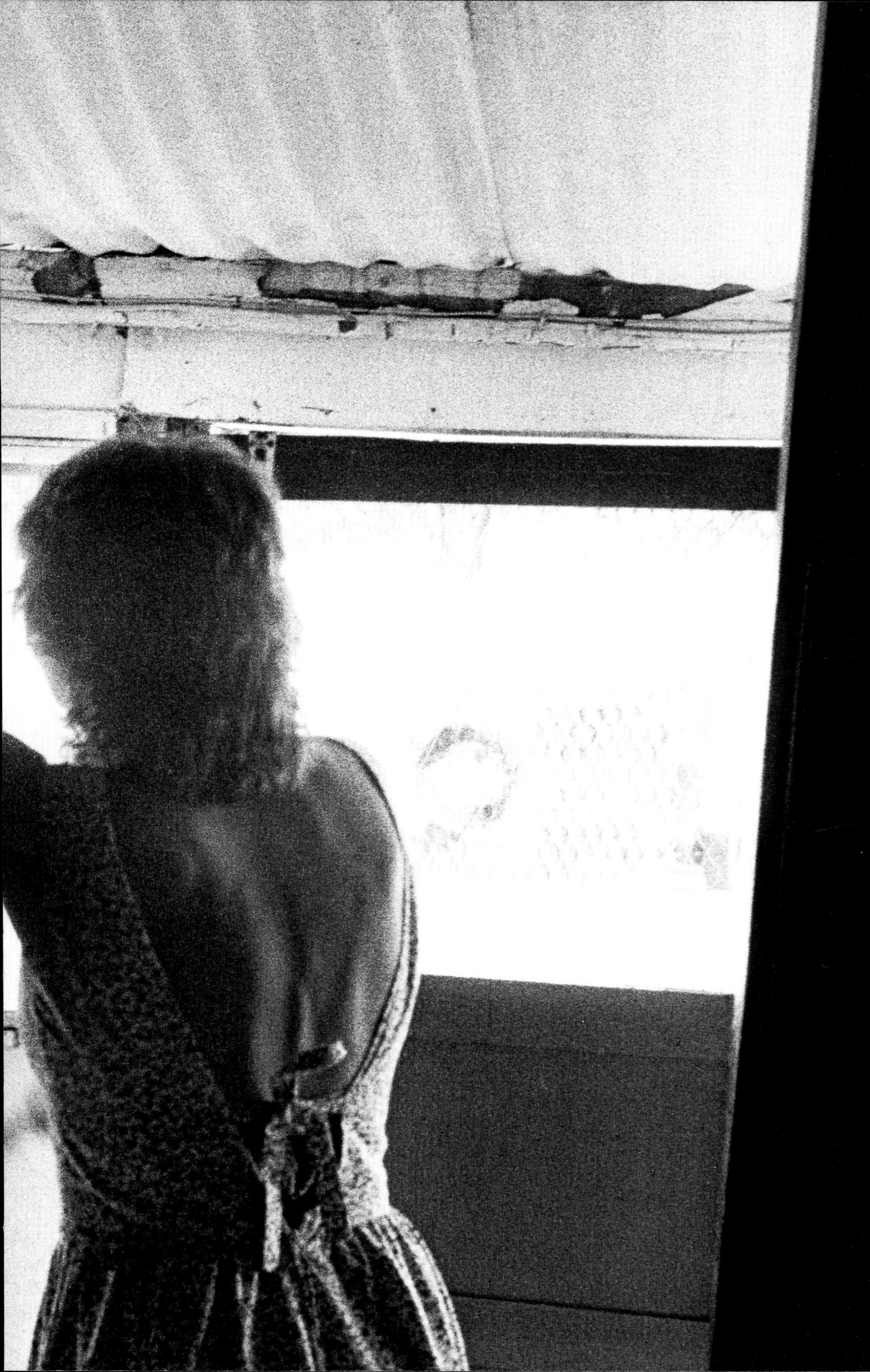

PEE WEES AT THE POINT

TROPPO SIGNIFICANT PROJECTS

1980
Coleman house, Fannie Bay
Wedd house, Milner
Lawler house, Coconut Grove

1981-82
Easterbrook house (unbuilt), Berry Springs
Low Cost house (Green Can), Karama
Elevated Green Can: Worth house and
    Buck house, Coconut Grove;
    Gartland house, Karama;
    Draper house, Leanyer
Gow/Royale house, Howard Springs
Locke verandah, Stuart Park
Troppo house, Coconut Grove
Brazier house, Leanyer
YHA Kakadu (unbuilt), Kakadu
Roebuck Bay Resort (unbuilt), Broome
YHA Hostel, Alice Springs
Carr house, Coconut Grove
Creswick house, Virginia
Collins/Fenbury house, Katherine
Butcher house, Virginia
St Andrews Lutheran School, Leanyer
Red Cross Blood Bank, Darwin

1983
Pitt house, Palmerston
Welke/Bonney house, Nightcliff
Spazzopan house, Coconut Grove
Kaiplinger house, Coconut Grove

1984
NT Parliament House Design Comp., Darwin
Phillips house, Howard Springs
Gerovich house, Parap

1985
De Graff house, Brinkin
Rossetto house, Northlakes

1986
Calland house, Humpty Doo
George house addition, Fannie Bay
Melville house, Humpty Doo
Rhodes house, Northlakes

1988
Lyne house, Coconut Grove
Dimon Project house
Bird Information Centre, Kakadu

1989
Dream House (unbuilt)
Addison house, Derby WA
Hazeldine house, Northlakes
Kakadu National Park Stage 3: Ranger Housing
Quinn/Walker house, Coconut Grove
Opium Creek Station Manager's house, (unbuilt)
    with Glenn Murcutt, Point Stuart

YHA Pioneer Hostel, Alice Springs
Rathie house, Northlakes
Price house additions, Fannie Bay
Wigram Island Resort (unbuilt),
    East Arnhemland
Katherine Aboriginal Cultural Centre (unbuilt),
    Katherine
Giblin house additions, Ludmilla
Gunlom ablution block, Kakadu
Rumah Kampung/Kerr house, Howard Springs

1990
Tropical house, Palmerston, with Speargrass
    Architects
Inside-Outside house, Milikapiti
Tom Harris donga, Humpty Doo
Red Cross Nursing Home, Katherine
Bakhita Village Group Home, Coconut Grove
Kakadu National Park: Northern Entry Stations
    and public toilets
Porcelli/Hazeldine house, Malak

1991
Marrkolidjban Outstation School, Central
    Arhhem Land
Farrant house, Gooseberry Hill WA
Christo's-on-the-Wharf, Darwin
Press/Taylor house additions, Coconut Grove
Maningrida house, Maningrida
DHA Project Houses, Darwin
Elix house, Nightcliff
Hall house addition, Coconut Grove

1992
Kakadu National Park Southern Entry Station
DHA Larrakeyah/Coonawarra Medium Density
    Housing Precincts, Darwin
Bowali Visitors Centre Kakadu National Park,
    Kakadu with Glenn Murcutt
Stokes St house, Parap

1993
Maloney house, Palmerston
DHA RAAF Infill Housing/Upgrades, Darwin
Gutjangan Outstation School, East Arnhem land
Arid Zone house, Alice Springs
Troppo house additions, Coconut Grove
NLC Regional Offices/Staff Housing, Top End
Vitori house, Humpty Doo

1994
Resort Xanadu (unbuilt), Port Douglas
Kaiplinger Units, Darwin
Barb's Love Shack, Milner
Breig house, Kununurra WA
Copeman house addition, Rapid Creek

1995
Saunder's verandah, Nightcliff
Kingsley-Khoo house, Berry Springs
Katherine Terrace Shopfront, Katherine
Video 2000 Shopfront, Katherine
Old Contractor's Camp, Jabiru

1996
Wyville Smith house, Coconut Grove
Rockfall Replacement Housing, Christmas Island
Esperance Holiday Apartments, Esperance
Nganampa Health Clinics and Staff Housing,
    Central Australia
Welke/Bonney house addition, Nightcliff

1997
Katherine Cross Roads Visitor Information
    Centre
Community Housing Projects: Kowanyama,
    Galiwin'ku, Rammingining, Kakadu,
    Mowanjum, Kandiwal/Larinuywar

1998
Top End Hotel Refurbishment, Darwin
Nganampa Aged Care Facility, Central Australia
PeeWees at the Point, Darwin
Thiel house, Cullen Bay

TROPPO STAFF - MAY 1999

(L TO R) ADRIAN WELKE (DIRECTOR), GLENN HOWE, JOANNA REES, ANDREW O'LOUGHLIN, FIONA HOGG, RICHARD LAYTON, PHIL HARRIS (DIRECTOR).

PAST COLLABORATORS AND LONG TERM STAFF MEMBERS INCLUDE:
PETER SAVAGE, CATHY BAILY, PETER TONKIN, SUE HARPER, GREG NORMAN, MICHAEL WELLS, DANNY WONG AND GLENN MURCUTT

OTHER PREVIOUS STAFF MEMBERS:
DON STELLARD, DAVID RUTTER, PETER MUSKINS, BEC FRANCIS, SIMON BROWN AND NICK LYMN.

BIBLIOGRAPHY

The Australian, June 25 1982
Architecture Australia, June 1983
The Australian, June 22-23 1985
Australian Built, Exhibition, NSW Gallery September 1985, and subsequent national tour
Sydney Morning Herald, December 19 1985
The Australian, February 22-23, 1986
Architecture Australia, May 1986
Architecture in Australia since 1960 Jennifer Taylor, Law Book Co, Sydney 1986
State of the Arts, ABC TV, July 1986
Architecture Australia, September 1987
Housing, Dwellings and Homes, Design, theory, research and practice, Roderick J Lawrence, John Lokey & Sons 1987
Building a Nation, A History of the Australian House, John Archer, Collins 1988
Made in Australia, A source book of all things Australian, William Heinemann Aust. 1986
Dream Homes Exhibition, Australian Centre for Contemporary Art, Melbourne, June 1988
Poetry in Iron, Philip Goad, Paola Tombesi, Riccardo Vannucci, in Spazio & Societa, No. 43 Jul/Sept 1988
Australia: verso un'architettura modestamente "galvo"?, Reyner Banham in Casabella, No. 550 Oct 1988
Technica e Prodotti: le coperture metalliche in Ville Giardini, No. 234 Feb 1989
Identifying Australian Architecture: Styles and Terms from 1788 to the Present, Apperley, Irving & Reynolds, Angus & Robertson, Sydney 1989
The Age, Melbourne, May 15 1990
Architecture Australia (Flightpath Architects), August 1990
The Age, October 30 1992
Architecture Australia, November 1992
Building Today, November 1992
House & Garden, April 1993
Business Review Weekly, April 2 1993
Lumen, (University of Adelaide), Spring 1993
The Australian, August 1993
Architecture Australia, October 1993
Mubibab, (Royal Brunei Airlines), October 1993
The Weekend Australian, July 30-31 1994
The Bulletin, August 9 1994
Arts Today, (SBS Television), September 25 1994
Corporate + Office Design, October 1994
Architecture Australia, Nov/Dec 1994
Architectural and Interior Specifier, December 1994
Territory Construction Journal, February 1995
Our House, (Channel 9 Television), March 9, 1995
Vogue Living, April/ May 1995
Architettura Contemporanea dal 1943 agli anni 1990, Corrado Gavinelli, Jaca Milan 1995
NT News, 14 June 1998
The Kalgoorlie Miner, August 1998
The Australian Magazine, 3 October 1998
NT News, 4 October 1998
Better Homes & Gardens, November 1998
Architecture Australia, Nov./ Dec. 1998
Home and Gardens, NT News, 10 November 1998
Building Australia, November 1998
Monument Yearbook, 1998
The Architectural Review, (UK) February 1999
NT News, 30 March 1999
Architecture Australia, March/April 1999
NT News, 18 May 1999
RAIA Chapter News, June/July 1999
The Esperance Express, 22 July 1999
NT News, 14 August 1999

PUBLICATIONS

Influences in Regional Architecture, A C Welke, J J Hill, J N Hayter, P N Harris Adelaide 1978 A study of regionalism in Australian architecture resulting from a tour around our Continent's perimeter
Darwin: a Map Guide to Architectural Heritage of the City, A C Welke & P N Harris for the Royal Institute of Architects (NT) Darwin 1981
Punkahs & Pith Helmets: Good Principles of Tropical House Design, P N Harris & A C Welke, Darwin 1982
(Also published in a format for schools by Education 1982) An investigation and attempt at definition of the Top End vernacular, eliciting principles still applicable in designing and building houses for today

PAPERS AND OTHER PUBLISHED WORKS

Relevant Housing: An Historic Overview of Tropical Housing in the NT with Implications for Future Solutions, P N Harris & A C Welke in Transactions of the Menzies Foundation Volume 2, 1981
Various newspaper articles for *The Star* Darwin 1981 and *The Sunday Territorian* Darwin 1984
Various letters and statements in relation to the Myilly Point Conservation Campaign 1983-94
Various radio interviews, letters and statements in relation to urban environment issues 1989-93

LECTURES, PUBLIC ADDRESSES

Relevant Housing: An Historic Overview of Tropical Housing in the NT with implications of Future Solutions, Menzies Foundation Conference, June 1981
The History of Regional Expression in Top End Housing, History Teachers' Conference, May 1982
Various addresses on *Tropical Housing* to year 11 class groups, 1982-90
Climate as a Determinant in Tropical Architecture, Geography Students' Conference, June 1984
Troppo Architects: Architecture is a Community Art, Lecture program through the Department of Architecture University of Queensland, March 1985
Troppo Architects: Architectural Jam, 1987 Biennial Oceanic Architectural Educational Congress Hobart, Tasmania, May 1987
A Hedonist's Handbook to Full Enjoyment of the Elements, ANZAAS Conference, Townsville, September 1987
Troppo Architects: Lecture, School of Architecture, University of Adelaide, October 1988
The Rise & Fall of the Darwin House, Under the Banyan Tree Guest Lecturer, State Reference Library, Darwin, February 1990
In Defence of the Tent in Defiance of the Cottage, The Australian Dwelling Conference Hay, NSW, May 1990
RAIA Architecture Polemic Lectures, Melbourne, September 1990
Architecture in Isolation Conference, Perth, October 1990
Seminar sponsored by University of Adelaide *Thermal Preferences in Housing in the Humid Tropics,* Adelaide, September 1990 Darwin, October 1990
Circus 1991 Biennial Oceanic Architectural Educational Congress, Brisbane, July 1991
Towards Developing a More Appropriate Built Form in Darwin, CUE Action Group Public Meeting, Darwin, October 1992
1993 Biennial Oceanic Architectural Educational Congress, Adelaide, April 1993
Troppo Architects, RAIA (ACT), November 1993
Tropical Housing Conference, Dept. of Lands, Planning and Local Govt./ DHA,Townsville, May 1994
Master Class, PNG University of Technology, September 1994
Bowali Visitor Centre for RAIA (NT) May 1995
Bowali Visitor Centre, inc detailing workshop, RAIA (NSW) June 1995
Professional Development Program, RAIA (SA) June 1995
Architecture Week Keynote Address, RAIA (WA), July 1995
Art and Architecture: Intersection Conference, Adelaide Festival of Arts, March 1996
Newcastle University Architecture Students Architecture Week, July 1996
Troppo Architects, RAIA (SA) Sessions, Boltz Cafe, Adelaide, August 1997
Public Architecture in the Top End, WA Shire Councils Assoc. Conference, Darwin, November 1998

AWARDS

Darwin Parliament House Competition, August 1984
 - Commendation
Tropical House Design Competition, Darwin NT, June 1990 - Winners
 (in assoc. with Speargrass Community Architecture)
Tropical House Design Competition, Darwin NT, June 1990 - Commendation
 (in assoc. with Speargrass Community Architecture)
RAIA (NT) Tracy Memorial Award, 1992
 - for Works in Kakadu for A.N.P.W.S.
RAIA (NT) Tracy Memorial Award, 1992
 - for Marrkolidjban Outstation School
RAIA (NT) Architectural Innovation Award, 1992
 – for Marrkolidjban Outstation School
RAIA (NT) Architectural Innovation Award, 1992
 - for Troppo Type 5 Tropical House (in association with Speargrass Community Architecture)
RAIA (NT) President's Award for Recycled Buildings, 1992
 – for Christo's on the Wharf
National ARIA Special Jury Award, 1992
 - for architecture in Northern Australia
Arid Zone House Design Competition, Alice Springs, 1992 - Second Prize
 (in association with Speargrass Community Architecture)
RAIA (NT) Tracy Memorial Award, 1993
 - for Medium Density Precinct 2, Larrakeyah
National RAIA Robin Boyd Award, 1993
 - for Medium Density Precinct 2, Larrakeyah
RAIA (NT) Tracy Memorial Award, 1994
 - for Bowali Visitor Centre, Kakadu National Park (in association with Glenn Murcutt + Assoc Pty Ltd)
RAIA (NT) 'People's Choice' Award, 1994
 - for Bowali Visitor Centre, Kakadu National Park (in association with Glenn Murcutt + Assoc Pty Ltd)
RAIA (NT) President's Award for Recycled Buildings, 1994
 – for Bakhita Centre Group Home
National RAIA Sir Zelman Cowen Award for Public Buildings, 1994
 – for Bowali Visitors Centre, Kakadu National Park (in association with Glenn Murcutt + Assoc Pty Ltd)
Metal Building Award, 1994-5
 - for Bowali Visitors Centre (Kakadu National Park) (National Award by BHP/Metal Building Products Manuf. Assoc.)
RAIA (NT) Burnett Award, 1995
 - for Porcelli/Hazeldine House
RAIA (NT) 'Peoples Choice' Award, 1995
 - for Porcelli/ Hazeldine House
RAIA (NT) Tracy Memorial Award, 1998
 - for Pee Wees at the Point, Restaurant
National RAIA Award for Commercial Buildings, 1998
 - for Pee Wees at the Point Restaurant
RAIA (NT) Burnett Award, 1999
 - for Thiel Residence
RAIA (NT) President's Award for Recycling 1999
 - for Top End Hotel Refurbishment

ADRIAN WELKE AND PHIL HARRIS

PHOTOGRAPHY CREDITS

All photography © by Patrick Bingham-Hall
except as listed below:

Troppo Architects © p14; 24; 26 top left, top right; 28 top left, centre right; 29 top right, bottom left; 30; 31 top left, bottom left, bottom right; 34 top left, bottom left, bottom right; 35 top left, centre left, bottom left, bottom right; 44 top left; 46 top left, centre right, bottom right; 48 centre right, bottom left; 49 top left, bottom left, top right; 50 top left, centre left, centre right, bottom; 51 bottom left, bottom right; 55 bottom right; 56 centre left, centre right, bottom; 57 centre right, bottom right; 64 top right; 67 centre right, bottom left.

John Gollings © p45; top, bottom; 59; 60 bottom left; 61 bottom left